Flowering: Physiological, Biochemical and Molecular Aspects

Flowering: Physiological, Biochemical and Molecular Aspects

Ajit Kumar Ghosh

CYBER TECH PUBLICATIONS
4264/3, Ansari Road, Daryaganj, New Delhi-110002 (India)
Ph.: 011-23244078, 011-43559448 Fax: 011-23280028
E-Mail: cyberpublicationsdelhi@yahoo.com
Website: www.cybertechpublications.com

FLOWERING: PHYSIOLOGICAL, BIOCHEMICAL AND MOLECULAR ASPECTS

Ajit Kumar Ghosh

First Edition 2010

Published by :

G.S. Rawat for **Cyber Tech Publications**
4264/3, Ansari Road, Daryaganj, New Delhi-110002 (India)
Ph.: 011-23244078, 011-43559448 Fax: 011-23280028
E-Mail: cyberpublicationsdelhi@yahoo.com
Website: www.cybertechpublications.com

Preface

The flowers of angiosperms have fascinated both humanists and naturalists for as long. Flowers are the source of love, affection, emotions, inspiration on one hand, while it is the prime source of fruits and seeds that make up the bulk of the world's food supply has given near-religious significance to flowering in many cultures.

Plants have evolved m hanisms by which leaf senesoence can be induced by many stresses to reallocate nutrients to reproductive organs and to eliminate water consumption by older, less productive leaves. This regulation of leaf senescence has an obvious adaptive value, allows in the plant to complete its life cycle even under stressful condition. Senescence is a correlative event and is the cumulative effect of the formation of flowers, fruits, seeds, apices and roots. Senescence is a mechanism, which works in favour of maintaining an ecological balance both for the internal milieu and the external environment of a species

This book provides a simplified description of the flowering at the molecular, biochemical, organelle, cell and organ levels of organization in plants, which contribute to the complex processes. The text is intended for post-graduate study and for non-specialists in other disciplines who wish to understand flowering mechanism and control. References are mainly from the secondary or review literature, from which detailed arguments and the primary literature may be obtained. Several recent textbooks have considered more specialized aspects of flowering and there are many excellent reviews; this book is intended to point the reads to them

—*Author*

Preface

CONTENTS

1

Introduction

Angiosperm embryology is the study of the floral structures and processes directly responsible for seeds. These include stamens (collectively the androecium) and pollen development; pistil or carpels (collectively the gynoecium) and embryo sac development; pollen germination, pollen tube growth and fertilization, and endosperm and embryo development (and sometimes the enclosing seed cover). One could also include seed dormancy, and even onset of seed germination. It is a discipline dominated by all forms of the microscope.

The emphasis is on normal sexual reproduction using economic plants (including garden flowers and weeds) as examples, but what is presented certainly applies to embryology of flowering plants in general. Apomixis and polyembryony, which are forms of asexual reproduction that use some of the sexual apparatus, are also mentioned briefly.

The total number of described angiosperm species is an elusive figure, with estimates ranging from 230,000 to somewhat over 400,000 (Govaerts, 2001). The subset that includes plants of economic significance (including weeds) is also subject to widely varying estimates, but it easily encompasses several thousand species. Although this book concentrates on the subset of economic plants, which has been neglected almost completely by previous embryology texts, it is obviously impossible to cover all such plants. If such were possible, the result would almost certainly be an unreadable encyclopedic tome. The goals of this book, in light of the necessary compromises just mentioned, are to:

- Present an outline of angiosperm embryology.
- Put flesh on the life cycle skeleton with supporting facts and enough specific examples to give some appreciation for general embryological features, as well as the diversity among economic plants.
- Present some research results, hypotheses, and speculations to try to explain why some things are as they are.

Every biological discipline is infinitely complex, which sustains the interest of professionals but can intimidate others, and embryology is no exception.

Most of the richness of detail about angiosperm embryology reported in its vast literature (easily 30,000 or more publications as an estimate) is therefore necessarily absent from this account, but entry into more specialized secondary and primary sources on most topics is provided by the many citations in the text that are listed at the end of each chapter.

My intention is that this book should hover, figuratively speaking, somewhere between a ground-level introductory text and the rarified upper air of a fully detailed research volume—not an easy task since there is no altimeter besides experience to gauge the proper elevation.

Two sections on background that follow describe the standard books on angiosperm embryology. I mention the number of references in the bibliographies of some of these works to convey an idea of research progress over time, and also to illustrate the huge problem of how to choose a necessarily limited number of publications to support a balanced narrative of reasonable length.

The last three sections of this chapter deal with the nature and evolutionary interpretation of the flower, introduce the components of a flower, and finally present a general diagram of the angiosperm sexual cycle, with a brief accompanying explanation. These last three sections set the stage for the nine more specific chapters that follow.

BACKGROUND: GENERAL WORKS ON EMBRYOLOGY

When the astounding discovery of "double fertilization" (two sperms needed for embryo and seed development—almost a unique characteristic of flowering plants) was made near the end of the 19th century, it suddenly fitted earlier pieces of knowledge into a coherent story. The study of angiosperm sexual reproduction using microscopy suddenly became a very popular endeavor, and the resulting spate of information soon required some organization. The first synthesis was by Coulter and Chamberlain (1903), in a book that is still useful and readable. They used the term "Special Morphology" for what later came to be called embryology. Their treatment omitted floral morphology, which mostly concerns form and diversity as revealed by the naked eye and the low-power magnification of a hand lens, and pollination biology. These two disciplines had already accumulated a substantial published literature for about 200 years by the end of the 19th century. Surprisingly, Coulter and Chamberlain also remarked in several places that published information about certain embryological topics had already exceeded what could be digested by reasonable effort, thereby acknowledging that use of the microscope for embryological studies had already burgeoned by the early 20th century. Thus the field of study that would soon be called embryology was already well launched.

Coulter and Chamberlain devoted most of their book (about 250 pages) to chapters on pollen and its development, the embryo sac, fertilization, endosperm, and the embryo. They utilized published information about any and all angiosperms without emphasizing economic plants. Their chapter organization has served to define the scope of embryology for later investigators. Their long section on embryology was really a book within a book, followed by about 100 pages dealing with the already well-defined disciplines of taxonomy, anatomy, paleobotany, and evolution.

A quarter of a century later, Schnarf (1929) devoted almost 700 pages to mostly the same embryological topics, but in more detail since he included information from the greatly increased number of published studies since the Coulter and Chamberlain book. Schnarf also added chapters dealing with apomixis and other asexual aberrations of the sexual cycle, of which there are many among angiosperms. Schnarf also ranged over all of the angiosperms, with no emphasis on economic species. His weighty volume was not written for easy reading, even for those who could read German.

Two decades later, P. Maheshwari (1950) published his classic text, mostly organized like Coulter and Chamberlain's chapters, but with the addition of an introductory historical chapter and some chapters on more specialized topics, including one on the emerging field of experimental embryology. Maheshwari also selected his information from among all angiosperms, with no particular attention to economic species. This treatment, like that of Coulter and Chamberlain, was reasonably readable and suitable for advanced undergraduates and graduate students.

After a lapse of 13 years Maheshwari in 1963 edited a multi-authored book that was really an expanded second edition of his 1950 work, and a tacit admission that knowledge appeared to have swelled beyond one person's ability to encompass it. Each chapter presented a detailed treatment of one topic, which elevated the book to advanced text status.

Continued accumulation of published work on all aspects of embryology, as well as the emergence of several specialties based on new technology and experimental work, required another grand overview after 20 more years. This appeared in the form of an 830-page multi-authored volume (Johri, 1984), which the preface described accurately as an "advanced treatise." This comprehensive work is presently the most current detailed general treatment of angiosperm embryology, that is, a treatment of mostly the same topics in the same order as included in earlier embryology works.

A new comprehensive multi-authored work is at hand, however, after another lapse of 20 years, this one in three volumes. Volume 1 is available (Batygina, 2002) as of this writing; it deals with topics up to the female gametophyte (embryo sac).

Another advanced treatise (Raghavan, 1997) is a one-person effort that covers pollen development, pollen and pollen tube interaction with the carpel, embryo sac development, fertilization, embryo, and endosperm, but with a strong emphasis on molecular, genetic, and in vitro studies, as reflected in the book's title, *Molecular Embryology of Flowering Plants*. The almost 700 pages include a 150-page bibliography of over 5000 references, most of which concern research trends of recent decades.

Other recent treatises concentrate on only a specific portion of the embryological cycle, and many of these are cited later in appropriate chapters. Although not a book, the review article of Prakash (1979) deserves mention here because of its general usefulness for those interested in economic plant embryology. He provided a list of 720 economically useful species in 89 families that had been studied embryologically up to that time, citing published studies for each species. Also, a recent multi-authored text deals with the general embryology of temperate zone woody fruit plants (Nyéki and Soltéz, 1996).

BACKGROUND: EMBRYOLOGY AND SYSTEMATICS

Already by the early 20th century enough work had been done to show that embryological structures and behavior vary enormously in detail among and within angiosperm families. It soon became evident that someone needed to organize such information for the benefit of taxonomy and systematics. The first such collation appeared in 1931, when Schnarf reorganized and expanded the information from his 1929 volume and presented it by family categories. He included over 1500 references in this illustrated treatment.

It took another 35 years before Davis (1966) published her exhaustive scholarly work based on information gleaned from a vast 202-page bibliography of more than 4,500 references. The bibliography itself was almost of book length, which forced Davis to squeeze the text to almost telegraphic conciseness and omit illustrations. There is an introductory general description of the kinds of embryological features significant for systematics, followed by descriptions of embryological features listed by families. More recently an even more ambitious 2- volume illustrated treatise has been published (Johri et al., 1992), also organized mostly by family chapters. There is a general embryological introduction of 112 pages, which amounts to a brief textbook by itself. The family accounts are based collectively on about 5,000 references, not all of them more recent than in Davis' account, but the new references carry the literature up to 1991.

A quite different and clever visual presentation of embryological information for dicot families and orders (monocots are omitted) is that of Dahlgren (1991). She depicted various embryological features as symbols superimposed on each of 23 repetitions of the same "bubble diagram" of the dicotyledons. Each circular-to-ovate bubble in the diagram represents one order and is of a size relative to the number of species included in it. The bubbles are arranged in their presumed evolutionary positions and distances from each other; thus the dicot orders are shown in their presumed natural relationships. These diagrams therefore comprise a chart of embryological features.

WHAT IS A FLOWER?

Embryology as described earlier in this chapter deals with floral structures and reproductive processes occurring within them. But step back and ask a broader question: what is a flower? Forests of paper and lakes of ink have been consumed by theorists who have proposed numerous hypotheses and speculations and by those who have defended or criticized them. Speculating on what is a flower and how it has evolved is fascinating for many people, but in this book only a venerable and widely accepted theory will be presented, one that most people find easy to understand, and which may even be proven correct some day.

The "foliar theory," which is over 200 years old, interprets the flower to be a specialized stem tip or side branch with compressed internodes and nodes with attached appendages that represent leaves that have evolved into modified appendages to serve reproductive purposes. These leaf-derived appendages are sepals, petals, stamens, and carpels, produced in ascending order from floral base to tip. Because these floral appendages have evolved in innumerable and often radically different ways among the more than 300 flowering plant families recognized today, the attractively simple idea of the flower as a greatly shortened branch with specialized leaves is often difficult, if not impossible, to discern. This has stimulated formulation of many alternate hypotheses about how the flower originated and evolved. Some of these concepts of the flower are presented, along with a defense of the foliar theory, in a very readable article by Eyde (1975). In this book the foliar theory will be assumed, and accordingly the floral appendages considered to be modified leaflike appendages.

THE FLORAL APPENDAGES

There are four types of floral appendages. Most economic species have all four in one flower, but one or more types of appendage may be absent from flowers of some species. A representative angiosperm flower that illustrates the four appendage types is the cherry blossom shown at the beginning. The most leaflike appendage in most flowers is the sepal (collective noun is calyx). The typically green sepals serve primarily as floral bud scales, developing early and enclosing and protecting the other appendages as they arise and develop. In some groups, however, sepals have also become adapted to entice pollinators by exhibiting colors other than green or by bearing nectaries. Nectaries are, however, not restricted to sepals; among different families they are known to occur on any of the other floral appendages, and in some groups even between appendages on the floral axis.

The next appendages in most flowers are the petals (collective noun is corolla), also usually recognizably leaflike, but adapted to entice pollinators by color and to act as "landing fields" for pollinators (they often have patches of roughened epidermal cells that make landing easy for insects). Some groups of plants lack petals, especially those in which wind pollination or self pollination occurs. Sepals and petals (collective noun is perianth) are not directly involved in embryological processes, and they will be almost entirely ignored elsewhere in this book. Stamens, the third set of appendages (collective noun is androecium), are specialized for pollen manufacture and usually do not appear to be leaflike, but they can be linked via intermediate forms in some groups to truly leaflike stamens in certain tropical families thought to be at the base of the angiosperm lineage.

Carpels (or pistils) are the ovule- and seedbearing appendages (collective noun is gynoecium), and they are the least leaflike and the most difficult to link to leafy precursors, although they can also be linked by intermediates to carpels that look like folded leaves in some tropical families. Flowers that have both an androecium and a gynoecium are bisexual or perfect, the most common condition, and species with only one or the other are unisexumatural or imperfect, as in kiwi fruit, hemp, and mulberry, for example.

THE SEXUAL LIFE CYCLE

Starting with the mature seed, the progression from vegetative to floral structures and events can be followed clockwise, finally returning to the seed. This life cycle diagram includes some details above an introductory text level but avoids almost all of the many variations that would obscure the fundamental features of the life cycle.It will become evident in these later chapters that each part of this generalized sexual life cycle can be expanded almost endlessly in complexity.

The major division of the life cycle into Sporophyte (2N—double set of chromosomes) and Gametophyte (N—single set of chromosomes) generations is usually described as the "alternation of generations," a characteristic of all sexually reproducing organisms.

In angiosperms the sporophyte (Greek, "sporebearing plant") generation is the visible plant, which greatly exceeds the two microscopic gametophytes (pollen grain and embryo sac) in size and longevity; in other groups of plants, such as ferns, mosses, and many algae, the gametophyte (Greek, "marriage-partner plant") generation may approach, equal, or even surpass the sporophyte in size and longevity. Reproduction in these non-flowering plant groups depends on at least a film of external water so that the ciliated or flagellated sperm can swim to the egg cell. Sexual reproduction in flowering plants is independent of the need for external water, and many unique structural and physiochemical features have evolved as a consequence, of which the most conspicuous is the release from ground-hugging so that the sexual cycle can occur at considerable distances in the air (gymnosperms had already developed this adaptation much earlier).

The shift from 2N sporophyte to N gametophyte occurs when meiosis is completed. In both anther and ovule this process of reducing chromosome number by half, which is really the removal of one whole set of chromosomes, results in four offspring cells. In the anther these cells are called microspores, and all four will develop into pollen grains (microgametophytes). In the ovule, the four meiotic products are called megaspores, and three of them will usually degenerate, leaving only one functional megaspore to enlarge and develop into the embryo sac (megagametophyte). Furthermore, there are many more microspore mother cells in an anther compared to a single megaspore mother cell in an ovule. The result of this great numerical difference is that a typical flower sends out an army of pollen grains to pursue one embryo sac; this needs emphasis here because the simplified cannot show this great disparity in numbers. They both eventually produce a pollen tube containing two sperm cells, but in one type (tri-cellular pollen) the sperm cells form before the pollen grain is shed from the anther, whereas in the other type (bi-cellular pollen) the generative cell does not divide by mitosis to produce the two sperm cells until after the pollen germinates and produces a pollen tube of a certain length. Other differences are also associated with these two pollen types.

An almost unique feature of angiosperm reproduction is the requirement for two sperm cells for successful fertilization instead of just one. One sperm has the expected task of combining with an egg cell, but a second sperm is needed to initiate endosperm, a tissue unique to angiosperms. Despite a few exceptions elsewhere (a similar if not identical double fertilization occurs in two gymnosperm groups—see discussion by Magallon and Sanderson, 2002), the two-sperm requirement is one of the defining characteristics of angiosperms.

Variations in the number and form of floral appendages are the basis for classification of angiosperms into taxonomic units from order down to species. These variations are relatively easy to see, but hidden within the flowers are endless variations in microscopic embryological features and processes. These are of great intrinsic interest for the embryologist, but some knowledge and appreciation of them is also needed by more practical-minded investigators whose purpose is to manipulate aspects of the life cycle for increased production or other agronomic or horticultural purposes.

2

Stamen and Androecium

The Latin word stamen means approximately "standing thread," a good descriptive term coined in the 17th century for the common type of stamen with a terminal swelling (the anther) supported on a slender, threadlike stalk (the filament), as in a cherry blossom. Stamens seem to be simple entities, but they are not, as this chapter demonstrates. Stamens are neglected as objects of study in embryology texts except for pollen development within the anther. There is one book, however (D'Arcy and Keating, 1994), that describes various aspects of stamens. This chapter examines the stamen without involving pollen development, which is the topic for the next two chapters.

The collective name for all stamens in a flower is androecium (Greek, meaning "male household"). In a typically organized flower the stamens are initiated just after the petals have appeared; they appear at first as small featureless primordia (bumps) spaced evenly around the flank of the dome-like floral apex. Slightly later in development, each stamen resembles a leaf primordium, not surprising since according to the foliar theory of the flower the stamen is a specialized leaf, an interpretation implicit in its technical name of microsporophyll (Latin for "leaf that produces microspores"—the cells that will become pollen). But beyond this early stage a typical stamen quickly acquires features that make it quite unleaflike.

The stamens of a flower may all be entirely separate from each other, variously attached to each other (connate), or attached to other floral parts (adnate). Some stamens may even become sterile "staminodes" that produce no pollen and in some plants even develop into nectaries. The many possible arrangements will not be considered further here because they are well described in any book on plant taxonomy. McGregor (1976) includes many fine drawings of flowers of cultivated species (some reproduced in this book), which depict most of the range of stamen number, arrangement, and form.

The number of stamens in a flower differs among and within families, although 4, 5, 10, or even more than 20 are found most commonly among cultivated dicots, and 3 or 6 among cultivated monocots. But there are exceptions, as in the extreme example of the reduced flower of the poinsettia, *Euphorbia pulcherrima* (dicot: Euphorbiaceae), which has only one stamen. The variation in numbers of stamens usually has no obvious explanation, but constraints of floral architecture and type of pollination mechanism are certainly involved.

STAMEN VARIATION IN REPRESENTATIVE FAMILIES

It would take many pages to describe the different forms and arrangements of stamens, even among families with cultivated members. The following brief list of 11 families with economically important species includes representative examples from the spectrum of stamen variation. A few technical descriptive terms are also introduced and explained. An important feature is how the filament is attached to the anther.

The filament apex usually merges with the base of the anther (basifixed) but in some plants—for example, lilies and the wind-pollinated cereal grasses—it connects instead about halfway up the outer side of the anther (dorsifixed), where it forms a flexible swivel joint that causes the anther to flutter and shake out pollen even in a slight breeze.

For each family listed below, one important cultivated species is mentioned:

Apiaceae (e.g., carrot): 5 dorsifixed stamens, bent over (inflexed) in bud.

Asteraceae (e.g., sunflower): 5 basifixed stamens, with anthers fused to form a circle around the style (syngenesious); anthers also have a sterile appendage extending from either apex or base, and they dehisce longitudinally toward the stigma (introrse).

Brassicaceae (e.g., mustard): 6 basifixed stamens, 4 longer than the other 2 (tetradynamous).

Cucurbitaceae (e.g., cucumber): 1–5 basifixed stamens, often united; anthers dehisce longitudinally on the side away from the stigma (extrorse).

Ericaceae (e.g., cranberry): 5–10 basifixed stamens, opening by an apical pore, which sometimes terminates a long tube.

Fabaceae (e.g., soybean): 10 basifixed stamens, all connate by a raised membraneous sheath (monadelphous) or 9 so fused and 1 standing alone (diadelphous).

Malvaceae (e.g., cotton): numerous stamens, with filaments arising from a common columnar sheath that surrounds the gynoecium.

Poaceae (e.g., wheat): 3 long, slender anthers, dorsifixed to the filaments with a swivel-joint connection (versatile).

Rosaceae (e.g., apple): numerous separate stamens, often dorsifixed, in several whorls on a hypanthium.

Rutaceae (e.g., orange): variable number of stamens, with thick filaments that often merge laterally to form two or more groups of fused stamens.

Solanaceae (e.g., tomato): 5 or 6 stamens adnate to petals, equal or unequal in length, or only 2–4 functional, the other stamens reduced to sterile staminodes; basifixed, sometimes dehiscing by an apical pore.

STAMEN ANATOMY

The structure of a representative anther is almost invariably illustrated in textbooks by the large lily anther as seen in cross section. Both the anther and its supporting filament are usually described as traversed by a single central vascular bundle surrounded by several layers of parenchyma cells. This central zone of the anther is called the "connective" because the pollen sacs are embedded around its periphery. The four pollen sacs become reduced to two near the time of anther dehiscence because the cellular septum that separates each pair degenerates late in development. Before this happens, the four separate mature sacs are filled with pollen and lined completely by a layer of cells called the tapetum. Moving outward from tapetum to the epidermis, there are from one to several parenchymatous "middle layers" (usually crushed during development), the endothecium (subepidermal layer that develops fibrous wall thickenings late in anther development), and the epidermis, which often disappears before anther maturity and exposes the endothecium.

It is widely accepted that a single unbranched vascular bundle (vein) traverses the filament and the connective of the anther, ending blindly near its apex, but this generalization is too simple. Hufford (1980), for example, provided excellent three-dimensional illustrations of more complex vein configurations from anthers of some common plants. In the crabapple (*Pyrus* species, Rosaceae), for example, the single vein in the filament branches at the base of the anther; the two descending branches flank the lower part of the anther lobes, and the ascending vascular strand increases in diameter and gradually frays out.

In further detail, the phloem of the ascending vein is seen to occupy a peripheral position. In chokecherry (*Prunus virginiana*), also Rosaceae, the filament is dorsifixed but the filament bundle upon entering the anther terminates its xylem and divides into four short phloem bundles, which extend to the end of the short connective. These two examples illustrate considerable diversity of vascular arrangement in anthers even within one family. Vascular separation in the anther into phloem strands that approach the pollen sacs can be interpreted as an adaptation for a more efficient nutrient supply during pollen development. Vascular tissue in the filament as well as in the anther is dominated by phloem, which usually ensheaths the xylem like insulation around a wire (Schmid, 1976). This configuration is common in stamens (and roots) but uncommon in stems and leaves, and is probably also adapted to supplying the increased nutrient requirements of developing pollen. Leinfellner (1956) studied vascular development in the lily stamen and showed that phloem is well developed even before any xylem forms.

Xylem is relatively insignificant in the stamen, and in some there is an actual xylem gap at the base of the stamen (Lersten and Wemple, 1966). Such gaps are widespread in the dicot family Polygonaceae, e.g., in buckwheat (*Fagopyrum esculentum*) and rhubarb (*Rheum rhaponticum*) (Vautier, 1949). Continuous xylem is probably unnecessary because stamens complete virtually all of their development while effectively sealed from water loss within the flower bud. Discontinuous xylem, either from a built-in gap or caused by stretching and disruption of xylem cells during filament elongation, impedes water flow to the exposed anther, which contributes to its rapid drying out and dehiscence for pollen release.

Stomata occur on many stamens, more commonly on anthers than on filaments. This is especially true in monocots, where stomata even occur on the reduced anther connective of cereal grasses such as oats (Bonnett, 1961). A survey by Kenda (1952) showed that stomata are less frequent on dicot stamens, and are even absent entirely from many species; they are, however, common in such families as Brassicaceae, Fabaceae, and Rosaceae. Schmid (1976) mentioned that guard cells of some stamens are able to open and close, but in others they remain permanently open. Heslop-Harrison et al. (1987) opined that open stomata in lily anthers are another adaptive feature to speed up anther desiccation and dehiscence for faster pollen release.

GROWTH OF THE STAMEN:THE ANTHER

Most stamens have similar stages of gross development, which Kunze (1979) showed in a survey of 56 species from 32 families. A stamen is visible first as a slightly flattened (bifacial) primordium, which is a tiny bump of cells that resembles a leaf primordium. But this early foliar similarity ends when swellings appear along each margin of the primordium, so that the axis of a slightly older anther appears sunken into a median groove. These lateral bulges indicate where the pollen sacs (microsporangia) will develop. The two lateral swellings each develop a median furrow, so that an outline of each of the four internal pollen sacs can be seen externally. The filament, however, remains short and inconspicuous until the anther is almost mature.

An example of Kunze's general pattern of anther development is tobacco (*Nicotiana tabacum*), in which Hill and Malmberg (1996) showed both external changes, using scanning electron microscopy, and internal tissue differentiation in thin sections of anthers at the same stages of development, over a period of about seven days. The tobacco stamen grew from initiation to almost 1.2 mm in length during this period, most of which comprised the anther, and at the end the pollen sacs were already entering meiosis. Selected photomicrographs show some of the characteristic external changes during this time period. An anther can increase tenfold or more in length while pollen is developing, for example in maize (Moss and Heslop-Harrison, 1967) and oats (Bonnett, 1961), but in lily the anther increases closer to twentyfold between pollen sac initiation and the mature anther (Gould and Lord, 1988). The anther is therefore not merely a static vessel within which pollen develops, but is itself a rapidly developing and enlarging structure during this process.

Gould and Lord (1988) observed external changes of developing anthers in intact lily flowers by measuring distances between charcoal markings on anthers 1.1 mm and longer. They also combined scanning electron microscopy with thin sections of anthers in order to correlate external and internal changes. Omitting details, they concluded that the lily anther grows by a series of intermittent spurts. In the young anther these consist of waves of cell division moving from tip to base, but in the older anther these are replaced by tip-to-base waves of cell elongation instead of cell division. In their view, "Anther growth is a nonsteady system, with growth centers constantly shifting." They regarded earlier reports that describe growth in terms of localized meristems as too narrow and rigid. Their study revealed a previously unknown aspect of stamen complexity.

GROWTH OF THE STAMEN: FILAMENT ELONGATION

In most plants the filament does not elongate until late in stamen development, after the anther is full size and pollen is mature, or almost so. It then thrusts the anther out of the flower in a short burst of growth that is extraordinary for a plant organ. In lily, for example, Heslop-Harrison et al. (1987) reported while the filament is between 5 mm and 4 cm in length it elongates exponentially; therefore, individual epidermal cells of the filament must elongate up to seven times their previous length to compensate during this extraordinary spurt. Internally, new vascular cells form continuously to replace those continually being stretched and disrupted. The slender lily filament should, theoretically, become flaccid during this phase, but these investigators found that potassium ions flow continuously into the filament from the floral receptacle, which they interpreted as a fast, sensitive method of osmotic regulation to help maintain turgor pressure.

Grasses provide examples of extraordinarily rapid filament elongation. Arber (1934, p. 158–159) summarized the observations of several earlier workers. In wheat, for example, a filament at anthesis elongates from about 2 mm to almost 10 mm in just 2–4 minutes, which pushes the anther rapidly out of the floret. Rye filaments have been recorded as elongating at the amazing rate of 1.6 mm per minute. Cugnac and Obaton (1935) timed filament elongation during anthesis in 12 common grass species. In 10–30 minutes, filaments elongated 50–120% in rye, timothy, and other forage grasses. They concluded, after examining filaments that were rendered transparent and mounted whole, that epidermal cells do the elongating and remain intact, but internal ground parenchyma and cells of the vascular bundle are stretched and disrupted.

Such rapid growth suggests that unusual activity must occur within filament cells. Hess and Morré (1978) studied elongating lily filaments, but when they examined individual filament cells using transmission electron microscopy, no significant changes were detected in either the number or distribution of major organelles. There is as yet no subcellular evidence to help explain how some filaments can elongate at such amazing rates for a plant structure.

Some experimenters have applied auxins and other growth regulators in attempts to discover the mechanism of filament elongation. Greyson and Tepfer (1967) found that removing part of the anther of the cultivated fennel flower,

Nigella hispanica (Ranunculaceae), inhibited filament elongation in direct proportion to how much was removed. Furthermore, gibberellic acid placed on the anther-less filament tip enhanced filament elongation. They concluded that the stimulus for filament elongation emanates at least in part from the anther.

Koevenig (1973) performed similar elongation experiments on a species of spiderflower (*Cleome*; Capparidaceae), which has stamens with very long filaments. He found that indoleacetic acid placed on the filament tip of a decapitated stamen allowed elongation to occur, thus the growth regulator substituted for the anther. In a later review, Koning (1983) concluded that several growth regulators are either individually or jointly effective in different plants.

These experiments indicate that the stimulus for filament elongation emanates from the anther in the form of a hormone or other growth regulator, but that it is not the same chemical stimulus in all plants. This is an intuitively attractive mechanism because, if true, it keeps the filament short and quiescent while pollen develops; then it stimulates it by a chemical signal from the anther when pollen reaches maturity.

Chemical changes in the whole developing stamen have usually been correlated with stages of pollen development, and most of these events will be taken up in Chapter 4. It is, however, relevant to mention here that starch is abundant in the young filament of lily (Heslop-Harrison et al., 1987), but during pollen development it is progressively converted to sugars. Pacini et al. (1986) detected starch in the filament and endothecium of young cherry stamens until shortly after meiosis, when it began to disappear.

Milyaeva and Tsinger (1968), however, reported that starch in anthers of sweet orange (*Citrus sinensis*) did not appear until the beginning of meiosis, after which it accumulated in all cells of the anther wall and connective until shortly after meiosis ended, when it gradually disappeared.

In an earlier study of cultivated *Lilium* and *Amaryllis*, Woycicki (1924) reported abundant starch, but only in cells of the connective. The evidence from analysis of whole *Petunia hybrida* anthers is that the amount of carbohydrates declines gradually during meiosis and thereafter (Linskens, 1973). These several studies point to starch as a common energy source that accumulates early in stamen development and is gradually used up in later stages, but evidently not in exactly the same way in all species.

ANTHER DEHISCENCE

Pollen forms inside the anther, which means that an opening of some kind must form to allow the mature grains to be shed. This opening is usually in the form of two longitudinal slits, one between each pair of pollen sacs. Each slit, called a stomium (Greek, meaning "mouth"), opens by the zipperlike separation of special cells that form a strip between two pollen sacs, but it does not have to open for the entire length of a pollen sac.

In the Solanaceae, for example, it opens full length in sweet pepper, but in tomato and potato anthers it opens less than a quarter of the way below the tip of the sac, so that the pollen emerges only through a short apical slit. In maize, a similar pore-like slit opens only to slightly below the apex of the anther (Cheng et al., 1979). This restricted opening probably prevents pollen from emerging until the wind is strong enough to agitate the anther sufficiently to shake out grains and carry them some distance from the parent plant (Kiesselbach, 1949).

Anthers that open only by means of circular apical pores are common in the dicot family Ericaceae—for example, in azaleas and rhododendrons. In cranberry and blueberry (also Ericaceae), the pore is at the tip of a long apical tube. These curious "salt shaker" types of anthers are common enough to be known from 400 genera of 65 families of angiosperms (Buchman and Hurley, 1978).

Horner and Wagner (1980) described the stomium of sweet pepper, *Capsicum annuum* (Solanaceae). They concluded that calcium is withdrawn from the cell walls of the stomium and stored as insoluble crystals, thus weakening the walls and facilitating separation of cells in the stomium.

Such strips of crystal-containing stomium cells, with presumably the same functions, occur in other Solanaceae, in Ericaceae, and probably in other families. Examples are discussed, under the collective name of "calcium oxalate package" or "resorption tissue," by D'Arcy and Keating (1994). Calcium removal appears to be a common mechanism, but certainly not the only one, for opening the stomium. The cell layer that is perhaps more directly responsible for opening either a stomium or an apical pore in most species is the "endothecium" (Greek, meaning approximately "inner case").

In most flowering plants, the endothecial cells develop fibrous wall thickenings on inner tangential and radial walls late in anther development. The fibrous thickenings appear to be composed of cellulose that is not lignified (De Fossard, 1969). Among cultivated plants, good examples have been shown in sweet cherry (Pacini et al., 1986), garden bean (Whatley, 1982), and peas and lentils (Biddle, 1979).

Keijzer (1987) reviewed in considerable detail previous published work describing and explaining how anthers dehisce in flowering plants. Differential drying within the anther wall has been the most commonly reported mechanism, which causes stomium cells to shrink more than other cells and exerts a physical stress that ruptures the stomium. Keijzer conducted experiments on anther dehiscence in several common species, from which he concluded that water evaporates from the affects the epidermis and endothecium. The pollen sac wall bends inward and disrupts the stomium, aided by tangential expansion of the epidermis; these events are followed by an outward bending of the locule wall because the outer tangential wall of the endothecial cells shrinks compared to the fixed inner tangential walls, which are supported by the wall thickenings.

The stomium therefore not only opens longitudinally, but it separates further by rolling outward at the sides. Keijzer's work shows that when the anther opens to release pollen, much more is involved than a simple rupture between cells.

Keijzer's conclusion that anther dehiscence is not a simple matter was verified by Bonner and Dickinson (1989), who followed the formation of cell layers involved in stomium formation in tomato (*Lycopersicon*, Solanaceae).

A post-meiosis tomato anther with young microspores shows intact septa and stomia between pairs of pollen sacs. The cellular ridge intruding into each pollen sac is called a "placentoid," a growth of unknown function (more in Chapter 3 about this). Within the nearly mature anther the septum separating two adjacent pollen sacs ruptures, in part possibly because crystals of calcium oxalate accumulate, indicating withdrawal of calcium from septal cell walls and thus weakening them.

Certain epidermal cells form a longitudinal line of small weak cells along which the stomium will open, and the anther wall folds inward (inflexes) along this line. Finally, wall thickenings appear in the endothecium, and a combination of dessication and differential cell growth causes the stomial slit to open near the apex of the mature anther.

EVOLUTION OF THE STAMEN

According to Eames (1961) the stamens of early flowering plants lacked a defined filament and anther, consisting instead of a leaflike lamina with two pairs of microsporangia (pollen sacs) embedded parallel to the axis just to the inside of either surface. In more advanced modern angiosperm families, including virtually all economic plants, the laminar portion has been reduced, usually to just the connective and a more-or-less slender filament, with the anther sacs evident as moreor- less symmetrical bulges around the connective.

Modern stamens therefore probably evolved from an originally leaflike form capable of both photosynthesis and pollen production into specialized pollen-producers with additional anatomical adaptations for various modes of pollination. Stamen evolution was considered in far greater detail more recently by Hufford (1994). He did not overturn the Eames hypothesis, but suggested that it is not firmly supported. Hufford found evidence to support an alternative hypothesis, that stamens already evolved considerably early in angiosperm evolution, possibly in response to the equally early diversity of pollination mechanisms. Stamen evolution remains an intriguing problem for speculation.

3

Pollen Development: Theme and Variations

INTRODUCTION TO POLLEN

The word pollen (Latin, meaning "fine flour") is a good descriptor. Showers of pollen in pine forests, for example, can be gathered up in containers, where they do resemble flour. But individual pollen grains are tiny, ranging in size from about 5 ìm–300 ìm in diameter, although most are 25–50 ìm. A few typical grains could fit on the period at the end of this sentence. Pollen is the common name. The technical name is "microgametophyte" (Greek, meaning "plant that producés small gametes"). The "small gametes" are the male sperms. A microgametophyte is therefore a sperm-producing gametophyte.

A pollen grain is a tiny, short-lived, non-photosynthetic haploid plant that consists of only two or three cells. Pollen grains of most economic plants, and of plants in general, live for only a few hours to a few days after leaving the anther (Dafni and Firmage, 2000), and successful ones never touch the ground. A pollen grain is therefore an ephemeral speck that could well be regarded as a tenuous link in the life cycle.

But the usually "vanishingly" small chance of success of an individual pollen grain is compensated for because most species produce pollen in lavish overabundance. In addition, a pollen grain has an advantage over an amphibious free-living gametophyte of a lower plant because it does not release the sperms into an external film of water to swim to their destination. This means that even though pollination is a remarkably wasteful process for most flowering plants, when a pollen grain does land on the right stigma it will be nurtured and guided within the maternal tissue, with a good chance for successful fertilization.

Most pollen grains have a thick wall composed of two major layers. The usually conspicuous outer layer is called the exine, which in most grains is interrupted by one or three (or more in some grains) small circular-to-elongate thin areas called apertures. The pollen tube will emerge through one of these at germination. A small number of species have a uniformly thin exine that lacks any special exit sites. Because there is only one pollen tube, one can ask why most pollen grains have more than one aperture. There is no definitive answer as yet, but Chapter 4 includes speculations.

The exine is composed mostly of a tough material called sporopollenin, which defies natural and human efforts to erode or dissolve it. This remarkably durable structure can persist for millions of years if buried under oxygenfree (anaerobic) conditions, and the standard methods used to prepare fossil or modern pollen for microscopic study all involve subjecting the grains to extremely harsh chemical treatment, which removes all but the exine. The diverse sculptured exine patterns of pollen of different groups of plants provide what might be called a "skeletonized fingerprint" for species identification. It can be fairly said that the pollen exine and the similar sculptured outer layer of spores of other plant groups comprise the chief subject matter for the discipline of palynology (the study of spores and pollen).

The inner pollen wall layer is the intine, and in contrast to the exine it is composed mostly of cellulose. The intine has its own complexities but it is at least structurally and architecturally simpler than the exine. Chapter 4 includes details about pollen wall structure and development. A consideration of pollen development must also include attention to the formation of the pollen sacs themselves, especially the tapetum, the important cell layer that completely lines the interior of each pollen sac, and which is involved with pollen development.

SUMMARY OF POLLEN DEVELOPMENT

The general life cycle includes a brief outline of pollen developmental stages as part of the life cycle. A diagram that is more specific excludes the surrounding anther tissue to show how a sporogenous cell can progress to a mature pollen grain. The differences that shows between dicots and monocots are generally true, but there are exceptions. The differences between pollen shed with two cells vs. pollen shed with three cells are not differences between dicots and monocots, but instead seem to be correlated with the evolutionary advancement of a family, or even with certain groups within a family. The numbers 1–8 below mark steps in the progression from diploid sporogenous cells to haploid mature pollen:

1. Sporogenous cells proliferate by mitosis to a certain final number, and then each cell secretes an isolating callose sheath around itself.
2. The microspore mother cells (mmc) undergo meiosis, and most monocots form a cell wall after each meiotic division, whereas most dicots do not form any internal walls yet.
3. Dicots after meiosis have a tetrad of haploid microspore nuclei in a common cytoplasm, which is a coenocyte (Greek, meaning "common cell") before "pinching off" four microspores by a furrowing process.
4. Callose dissolves, releasing microspores into the fluid environment of the pollen sac.
5. Microspores enlarge, become vacuolate (i.e. deplete their food reserves and absorb water), start forming the exine, and then germinate internally to become bi-celled pollen.
6. The vacuolate pollen now has a large vegetative cell and a small generative cell.
7. The generative cell migrates within the vegetative cell, food reserves begin to accumulate, and the pollen may be shed at this two-celled stage.
8. In some groups the generative cell divides mitotically and produces two sperm cells (becomes tri-celled) before

the vegetative cell completes its engorgement with food reserves.

Descriptions of five actual examples follow. None of them illustrate all stages of pollen development; however, they complement each other. In addition, they provide glimpses of a few of the many variations in structure and processes that are known. A more detailed review of the processes that occur within anthers of angiosperms in general is that of Pacini (2000).

POLLEN DEVELOPMENT IN SORGHUM

Sorghum pollen development is representative of cereal grasses, but it also illustrates general features. The four pollen sacs develop more or less synchronously, but for ease of presentation each sac is shown diagrammatically at a different stage is smallest and shows the pre-meiosis situation: a central column of enlarged sporogenous cells is surrounded directly by the tapetum, a middle layer, subepidermis, and externally by the epidermis. The tapetum in grasses remains in place throughout pollen development (also the most common configuration among all angiosperms); therefore, it is called a "parietal tapetum" (Latin, "relating to the wall"). The functional term "secretory tapetum" is often applied to it because earlier investigators were sure that it secreted substances necessary for pollen development. In most grasses every sporogenous cell is appressed to the tapetum and remains so during all stages of development until the mature pollen is released from the anther. A recent survey of pollen arrangement has verified this unusual, perhaps unique, physical relationship for most grasses.

The enlarged single sporogenous cell at the right of pollen will undergo one or more mitoses before divisions cease, at which time the terminal number of sporogenous cells become microspore mother cells (mmc's). Just before meiosis begins, each mmc secretes callose between the cell membrane and the thin primary cell wall (next single cell). Callose is a gelatinous carbohydrate that isolates mmc's from each other and from surrounding diploid tissues during and after meiosis, when the transition to the haploid microgametophyte stage occurs. Most of the callose in a grass is deposited toward the center of the pollen sac, and only an extremely thin layer of callose lies between tapetum and mmc's. This uneven callose deposition helps to keep the mmc's appressed to the tapetum.

The anther with its four internal pollen sacs enlarges during meiosis. After the first meiotic division a new cell wall forms, dividing the mmc into two cells, which is called a dyad. After the second meiotic division two new cell walls partition each dyad cell into two new cells. This progressive wall formation is called "successive cytokinesis." The four haploid microspore cells remain attached to each other, forming a tetrad of microspores. The grass tetrad is unusual because it forms in one plane. Imagine a grass mmc as a pie cut in half by the first cytokinesis; then each dyad cell bisected again by the second cytokinesis, yielding four microspore pie quarters. This tetrad pie remains appressed to the parietal tapetum, and in this way each microspore maintains physical contact with the tapetum, an arrangement shown in pollen sac and in the next individual figure.

Meiosis and successive cytokinesis occur within the callosic sheath, which persists around the microspore tetrad for some time after meiosis. The exine layer of the pollen wall is now initiated around each microspore. Microspores enlarge as water is imbibed and stored in vacuoles. This enlargement occurs in concert with pollen sac and anther enlargement.

The microspore tetrads are surrounded, except where they contact the tapetum, by a fluid of largely unknown composition. Soon the callose that has encapsulated the microspore tetrad dissolves and contributes to the pollen sac fluid. Pollen sac is at this stage. As microspore vacuoles continue to absorb water and swell, the future pollen wall continues to thicken. While a microspore is enlarging, this wall must retain flexibility to accommodate its expansion, except where the single aperture, found in all grasses, forms along the surface of contact with the tapetum.

The several small microspore vacuoles expand and eventually merge into one large vacuole, an expansion that occurs as food reserves are used for metabolic purposes and converted into components of the developing wall. At this late vacuolate stage the microspore undergoes mitosis and becomes a pollen grain.

The distinction between microspore and pollen is frequently ignored, and in many published studies all stages after meiosis are called pollen. But there are good reasons to call the one-celled direct product of meiosis a microspore, and to restrict the term pollen grain to the two-celled structure resulting from microspore mitosis.

During the transition from pollen sac 3 to 4 the microspore divides mitotically, producing a large vacuolate vegetative cell and a tiny, lens-shaped generative cell. The nucleus of the vegetative cell migrates toward the pore, followed by the generative cell, which can now move because the thin callose sheath that briefly attached it to the vegetative cell membrane has dissolved. The tiny generative cell appears to float more or less freely within the vegetative cell, but in reality its movements are probably controlled by microtubules, tiny subcellular filaments that have been described from pollen of some plants.

New food reserves in the form of starch appear in the vacuolated pollen grain, and the central vacuole gradually shrinks. As starch accumulates, the tapetum gradually empties itself and collapses in place, and the pollen wall is completed. There is evidence from other species that nutrients enter pollen from the tapetum, and it can be reasonably speculated that the intimate pore-to-tapetum connection in grasses provides an efficient nutrient entryway.

While engorgement is proceeding, the generative cell divides to produce two sperm cells, a characteristic of grasses and several other families. Pollen sac 4 is mature, as is the last individual figure of the engorged pollen grain with two sperm cells. During the last maturation events in pollen, wall thickenings in the form of bars appear in the cell wall of the endothecium, the subepidermis of the anther. As explained earlier, these bars help the anther to dehisce and release pollen.

A detailed view of mature sorghum pollen, still with the single pore appressed to the now-collapsed tapetum, is 28 Flowering Plant Embryology, and a scanning electron micrograph shows a grain at the time of anther dehiscence.

POLLEN DEVELOPMENT IN SWEET PEPPER

Sweet pepper (Capsicum annuum, Solanaceae), is an herbaceous dicot. Its pollen development is similar to that of other members of this family—for example, the tomato (Sawhney and Bhadula, 1988)—as well as to many other dicots. Horner and Wagner (1980) illustrated the anther of sweet pepper with one of the four pollen sacs eliminated to show that there are three different types of crystals in the anther, each with its own characteristic distribution. One crystal type (druse) is associated only with the stomium.

Each pollen sac is a cylinder deformed by a longitudinal intrusion of connective parenchyma of unknown function, called a "placentoid," and each sac therefore appears U-shaped in cross-sectional view. A placentoid is common in the Gamopetalae (this term means "joined petals"), a group of related dicot families (including Solanaceae—for the placentoid in tomato pollen sacs) with fused petals that give many of the flowers the appearance of a trumpet bell. Within the placentoidal pollen sac of sweet pepper, stages of pollen development were shown in cross-sectional view by Horner and Rogers (1974).

Sporogenous cells form only two layers in the pollen sac; therefore, each sporogenous cell is in physical contact with the parietal tapetum. Tapetal cells toward the outside of the pollen sac (at top) are smaller than tapetal cells that line the inner placentoid-facing connective (at bottom). Here and in some other plants, it has been speculated without any evidence that the enlarged tapetal cells are larger because they are closer to vascular bundles in the connective and therefore more active in transfer of nutrients.

Callose is secreted by each mmc early in prophase I of meiosis; unlike cereal grasses (see sorghum example), callose in sweet pepper and in dicots in general is deposited quite uniformly between cell membrane and primary wall. Following meiosis, four microspores form, but by "simultaneous cytokinesis" instead of the simpler pie-slicing" successive cytokinesis described earlier for sorghum. The tetrad of microspores remains surrounded by callose, but when each microspore initiates its future pollen wall, the tapetum secretes an enzyme that dissolves the callose. The inner tangential and radial walls of the tapetum also dissolve away at about this time, leaving the partially naked tapetal cells somewhat deformed, like partly deflated basketballs. The microspores released from callose become immersed in the fluid of the pollen sac, which contains several dissolved substances. There is little information about this fluid, but polysaccharides have been reported by Gori (1982) in garlic, Allium sativa, and pectins in lily (Aouali et al., 2001).

The enlarging microspore wall continues to form while the cytoplasm gradually becomes more lightly staining as vacuoles accumulate and food reserves are steadily depleted. Microspores at the late vacuolate stage have a noticeably thicker wall and appear almost empty because of their large central vacuole. The tapetum in this figure shows signs of degeneration. Following microspore mitosis the pollen grain has a small generative cell within a large vegetative cell, food reserves begin to accumulate, and the parietal tapetal cells gradually collapse. When the pollen grain is mature and fully engorged with food reserves, the tapetum has degenerated almost completely and wall thickenings on the endothecial cells indicate that the anther is ready to shed pollen.

POLLEN DEVELOPMENT IN WALNUT

The walnut tree (Juglans regia, Juglandaceae) is a woody dicot in which pollen development conforms mostly to the general pattern shown in the previous two herbaceous plant examples. But walnut anthers form during one year, reach a certain pre-meiotic stage, and remain dormant over winter, completing their development during the following growing season.

This behavior occurs in many temperate zone trees and perennial herbs, particularly among bulbous monocots. The developmental stages described here from Luza and Polito (1988) include certain features that differ from the previous examples.

By the time sporogenous tissue and the layers external to it can be distinguished, the sporogenous cells form a multicellular column in each pollen sac. There are a large number of sporogenous cells; therefore, many of them do not contact the tapetum, unlike the cell configuration in pollen sacs of sorghum and sweet pepper. But the external anther layers are similar to those of the previous examples, consisting of tapetum, three middle wall layers, the endothecium (subepidermis), and the epidermis.

After the sporogenous cells complete their last mitotic division and become mmc's, each secretes a conspicuous callose deposit that becomes strikingly evident when viewed using a technique that causes the callose to fluoresce brightly. Meiosis without cytokinesis occurs while each mmc is encased in callose, and the resulting four haploid microspore nuclei spend a short period as a coenocyte. Simultaneous cytokinesis follows, which gradually pinches off the four microspores from each other by progressively invaginating constrictions, or furrows. The tetrad of separate microspores remains encased in callose for some time after meiosis.

As in the earlier examples of pollen development, callose dissolves and releases the four microspores of each tetrad into the pollen sac, where they gradually enlarge and become vacuolate as their food reserves are utilized to thicken the pollen wall.

The late vacuolate stage, and it also shows the typical collapsed appearance of the declining parietal tapetum. Microspore mitosis occurs during the late vacuolated state and initiates the twocelled pollen grain, which then begins to replenish its food reserves as the wall approaches maturity. The mature walnut pollen sac contains fully engorged pollen and a now completely collapsed tapetum. Thickenings appear in cells of the endothecium, indicating that the anther is ready to split open.

POLLEN DEVELOPMENT IN THE MUSTARD FAMILY

Cultivated members of the herbaceous dicot mustard family (Brassicaceae) have a representative dicot pollen development, but they also provide good examples of a widespread tapetal function among insect-pollinated species. The tapetum ruptures, releasing an oily substance called tryphine or pollenkitt, which enters the pollen sac and is deposited on and within the pollen exine. This tapetal coat makes pollen grains sticky, adhering them to each other as well as to pollinating insects and the stigma.

Tryphine includes carotenoid and flavonoid pigments, as well as proteins that will act as molecular recognition signals when pollen lands on the stigma. The interaction of these proteins on the exine with recognition proteins produced by the stigma is one of the gatekeeping mechanisms by which the stigma can reject foreign pollen and accept and stimulate germination of compatible pollen. The example discussed and illustrated here is a combination of a study of rape, Brassica napus (Grant et al., 1986), and radish, Raphanus sativa (Dickinson and Lewis, 1973).

A cross-sectional view of one pollen sac at the mmc stage shows that here, as in walnut, there are some mmc's deep within the central column that do not touch the tapetum. Also, it is evident that the very darkly staining parietal tapetum is quite conspicuous even at an early stage.

The tapetum remains turgid and conspicuous through some of the later stages, for example at the early vacuolate microspore stage. After meiosis the microspores start to fill up with food reserves even before they undergo mitosis to become pollen, another difference from previous examples. At the same time the tapetal cells begin accumulating dark lipid droplets; these droplets and other tapetal components are excreted from the tapetum and deposited both on the surface of the sculptured exine and within its many cavities.

A closer look at this excretion can be seen in a transmission electron microscope view of a radish anther at about the same stage. The microspores have a well-developed wall with many cavities, and the dark-stained excreted tapetal material can be resolved into two components, lipid droplets and larger vesicles with fibrous contents.

Somewhat later in development the tapetal cell membranes rupture, sending tapetal cytoplasm into the pollen sac, where some of it flows onto the surface and into cavities in the exine of the pollen wall. The tapetum here, and in other plants where it ruptures and contributes much of its cytoplasm to the pollen surface, is especially intimately involved in pollen development. The next example also emphasizes tapetal involvement.

POLLEN DEVELOPMENT IN SUNFLOWER

Sunflower (Helianthus annuus, Asteraceae) is an example of a startlingly different pattern of tapetal behavior as compared to the previous examples. All members of this large dicot family examined to date exhibit an "invasive" tapetum (Lersten and Curtis, 1990).

Invasive tapetal cells do not remain at the periphery of the pollen sac; instead, their cell walls dissolve at a certain stage and the tapetum collectively flows amoeba-like into the pollen sac and engulfs the developing microspores. As this unexpected process begins, the individual tapetal cell membranes merge to form a "syncytium," a single protoplasmic entity consisting of the merged tapetal cells.

The invasive tapetum is also called a "plasmodial tapetum," which calls to mind the syncytial stage of a slime mold. An invasive tapetum is common among monocots but uncommon in dicots except for the large family Asteraceae.

Pollen development in sunflower was described by Horner (1977). The early stages resemble those seen in the previous dicot examples, so the developmental sequence can be taken up after meiosis, starting with microspore tetrads encased in callose.

Up to this stage the tapetum has remained peripheral, but somewhat later, when the callose begins to dissolve from around the microspores, the tapetal cells exhibit irregular swelling and begin to lose their wall. Tapetal enlargement continues after the microspores are released by callose dissolution, and by the mid-vacuolate microspore stage the tapetal cells have elongated radially and are intruding deeply into the pollen sac, some now thrusting between the microspores. The late vacuolate microspores eventually become engulfed completely by the invasive tapetum.

While the microspores are still engulfed they undergo mitosis to form the two-celled pollen grain; at the same time the tapetum begins to develop vacuoles and show other signs of degeneration. While the pollen grains accumulate food reserves and the central vacuole gradually disappears, the tapetum declines further. It finally disappears completely, probably by absorption into the pollen, by the time the mature pollen is ready to be shed.

4

Pollen Development: Details of Stages

ANTHER DIFFERENTIATION BEFORE MEIOSIS

The tapetum is considered to be part of the pollen sac. Between the tapetum and the anther surface is the "anther wall" composed of the epidermis and some underlying cell layers. The developing anther wall can partition itself into its constituent layers in more than one way, which convinced some that distinctive anther wall "types" are characteristic of certain groups of plants. Davis (1966), for example, recognized four types of anther wall formation. But these are not mature static structures, and are therefore verifiable only by developmental study using statistically valid samples, which has seldom been done. A survey by Brunkener (1975) of early anther development that included several samples each of 60 genera from 38 angiosperm families, showed that individual anthers often deviate from their supposed types.

Even Coulter and Chamberlain (1903) commented early that local physiological conditions within the young anther could affect patterns of cell division and influence the developmental pathway. However it organizes itself, by the sporogenous cell stage the anther wall consists typically of epidermis, subepidermis (endothecium), and 1–5 intermediate wall layers.

Turning to the opposite, inward facing part of a pollen sac, the tapetum here abuts on a few to several parenchyma layers in which are embedded one or more vascular bundles; this central core of the anther is called the connective. Some investigators have proposed that tapetal cells next to the anther wall (the outer tapetum) have a different origin from tapetal cells facing the connective (the inner tapetum), and it is true that in some families the inner tapetum is composed of two layers of cells that are often larger in size. Whether tapetal location makes any physiological difference is not known, although the inner tapetum is obviously closer to the vasculature and therefore possibly receives more nutrients, or receives them earlier.

POLLEN SAC BEFORE MEIOSIS

Each pollen sac first becomes evident as a column of enlarged cells called the archesporium (Greek for "first cells to be sporelike"). Archesporial cells are also called sporogenous cells ("spore generators") because they proliferate by mitosis until a certain number have been produced. Mitosis then ceases and these final cells, which now become microspore mother cells (mmc's), will switch over to a final division by meiosis. Mitotic divisions in the proliferating archesporium may or may not occur in synchrony; in lily, for example, mitoses are scattered irregularly throughout (Walters, 1980, 1985). In wheat, lily, and barley, successive cycles of mitoses occur, each cycle proceeding more slowly than the one before (Bennett, 1984). In wheat, for example, the mitotic cycle that increased 12 cells to 25 cells took 25 hours, from 25 to 50 cells took 35 hours, and 55 hours were required to go from 50 to 100 cells. These successively slower mitotic cycles are not linked to any obvious cause, but they suggest that shifting from mitosis to meiosis requires a physiological transition period of as yet unknown nature.

The final number of archesporial cells (now the mmc's) within a pollen sac is consistent within a narrow range for an individual species but enormously variable among species. The range is from a single vertical row, as in lettuce (Lactuca sativa, Asteraceae) with only about 15 mmc's per pollen sac (60 total for an entire anther), to hundreds or even thousands per pollen sac, of which only those at the periphery contact the tapetum.

A single lily anther, for example, contains about 20,000 mmc's or 5,000 per pollen sac (Walters, 1985). In walnut it is obvious that many mmc's have no tapetal contact. All of the sporogenous cells of an anther are, however, connected to neighboring cells by numerous plasmodesmata. The great range in numbers of mmc's per pollen sac, and the differences in their physical relation to the tapetum, add to the difficulty of explaining what happens during the transition from mitosis to meiosis.

MEIOSIS

The Greek word meiosis means "diminution," which in a pollen sac refers to the reduction of chromosome number by half during the formation of four haploid microspores from one diploid mmc. This is not simply a separation of the original 2N sets of chromosomes to return to the N number of the original egg and sperm, because during meiosis there occurs crossing over and exchange of parts of paired chromosomes, which is the basis for genetic variation. There are several definitions of when meiosis begins (Bennett, 1984) and no agreement on what causes it to begin. One treatise on meiosis (John, 1990) states "The precise factors responsible for the transition from mitosis to meiosis are still largely unknown," a conclusion reiterated more recently by Li and Johnston (1999) and Shivanna (2003). Walters (1985) speculated that unidentified "meiosis-inducing substances" accumulate in the cytoplasm of pre-meiotic sporogenous cells, and when they reach some unknown critical concentration, mitosis cannot continue and meiosis begins. Indirect evidence for such progressive accumulation comes from the earlier-mentioned observation that sporogenous cells of wheat anthers take a longer period of time to complete each successive cycle of mitosis.

The progress of meiosis is marked by changes in appearance and behavior of the chromosomes, which provide convenient visible boundaries to define and identify stages, beginning with the onset of leptotene of prophase I. These stages are well known and widely described; here it should be a sufficient refresher to merely illustrate some of them from barley, Hordeum vulgare. Other events, less well known, will receive more emphasis. Each mmc secretes callose (1-3-ß glucan), a carbohydrate composed of linked glucose sugar units (Kauss, 1996), between its cell membrane and its primary wall.

Callose is referred to in older studies as "the special wall" and in some respects it acts like a temporary wall. Callose secretion coincides with, or may even precede, the onset of meiosis. Callose is usually deposited uniformly around a mmc, although it differs in thickness among species. In cereal grasses, however, it is distributed quite unevenly, extremely thinly between mmc and tapetum but deposited copiously in the center of the pollen sac. Fresh unstained callose viewed microscopically appears as a rather translucent gelatin-like substance.

Callose is a common secretion from various types of cells in all plant organs. Wherever it occurs it acts as a protective seal that prevents the passage of most molecules, but nowhere is it deposited more abundantly than in the pollen sac, where it appears to isolate mmc's both from each other and from the tapetum. Callose does not sever all cytoplasmic connections, however; some surviving plasmodesmata have been reported to even increase in diameter during meiosis into cytomictic channels. Such passages have been regarded by some investigators as necessary to maintain synchrony of meiosis. But just as mitosis in the premeiotic archesporium may be asynchronous, so mmc's do not always behave in concert. Walters (1985) speculated that mmc's enter meiosis at different times because they do not accumulate the "meiosis-inducing substances" uniformly.

The still unknown stimulus that initiates meiosis does not spread throughout a pollen sac the same way in all species. Although there is considerable meiotic asynchrony among lily mmc's (Shull and Menzel, 1977; Walters, 1980), Bennett et al. (1971) reported that meiotic stages within wheat and triticale pollen sacs always started at the base and proceeded toward the tip; in rye anthers the meiotic gradient was the reverse. In a survey of meiotic patterns in species of dicots and monocots, Neumann (1963) found gradients from base to tip, tip to base, and middle to both ends. Furthermore, many investigators have observed that different pollen sacs in the same anther are usually at somewhat different stages of meiosis. These many variations seem to defeat the formulation of a neat generalization.

During prophase of the first meiotic division there is a decrease in both ribosomal RNA and messenger RNA levels. Although these decreases have not been detected in all of the few species examined, some RNA reduction is expected because removal of the sporophytic genetic coding machinery has been postulated to be necessary before repopulation with gametophytic RNA can occur in the microspores. Mitochondria and plastids, however, do not decrease in number during meiosis, but they reportedly dedifferentiate to a simpler form that lacks the characteristic substructure of the mature organelles. According to Dickinson (1987) they can still synthesize DNA, and they often cluster closely around the nucleus, which suggests that these organelles have a functional relationship with the nucleus during meiosis. The initiation of, and causal mechanism for, meiosis remains mysterious, but following the protracted prophase of the first meiotic division, all subsequent stages appear similar to those of mitosis.

CYTOKINESIS

The process that cleaves the mmc into separate microspore cells after two or four microspore nuclei have been produced by meiosis is called "cytokinesis" (Greek for "cell motion"). It can occur in two fundamentally different ways. The first is by formation of a cell plate after each meiotic division, which is called "successive cytokinesis." This is really the same as cell plate formation following mitosis in vegetative cells, so one could reasonably expect it to also occur during meiosis. But contrary to expectation it occurs in only 40 families, of which 35 are monocots (Davis, 1966). Successive cytokinesis is easy to visualize; and the sorghum series of illustrations in Chapter 3 are representative.

It is far more common for meiosis to produce all four microspore nuclei before any walls form, which creates a temporary 4-nucleate coenocyte (cell with multiple nuclei). Each microspore nucleus, along with a proportional amount of surrounding cytoplasm, is isolated later into a separate cell by a process of progressive furrowing called "simultaneous cytokinesis." This unique mode of cell formation in plants occurs in 186 families, of which 176 (almost 95%) are dicots (Davis, 1966). The two types of cytokinesis are therefore mostly segregated between monocot and dicot lineages. Only four dicot and four monocot families have been reported to have species exhibiting both types of cytokinesis.

Simultaneous cytokinesis is not as easy to visualize in three dimensions because the furrowing is not a simple two-dimensional quartering process. Rather than a prose description, furrowing is illustrated, first from a species of Desmodium, a legume with pollen development identical to that of the common cultivated legumes. Following meiosis, the coenocytic tetrad is still surrounded by callose. Furrowing (cleavage) begins between each pair of nuclei, progressing until microspores are almost separated, except for slender connecting strands of cytoplasm. Furrowing is eventually completed, but the four separated microspores are retained for some time within the original callose sheath.

A second illustration of simultaneous cytokinesis is a more intimate one that follows the organized framework of two sets of microtubules that control certain intracellular movements. After the second meiotic division, the first set of microtubules links all four microspore nuclei, positioning them precisely within the coenocyte. A second set of microtubules later radiates from each nucleus, and cytokinesis proceeds along planes marked by the interaction of opposing arrays of these two sets of microtubules.

The behavior of these microtubules in relation to the microspore nuclei and the planes of cleavage was shown beautifully by Brown and Lemmon (1988) in the honeysuckle shrub (Lonicera japonica, Caprifoliaceae), and in Impatiens (Balsaminaceae). It is generally agreed that simultaneous cytokinesis has evolved from successive cytokinesis (Davis, 1966), which certainly seems a logical progression. But why should the most common mode of cytokinesis occur by furrowing instead of cell plates? The only suggestion has been that the triradiate (approximately pyramid-like) configuration of the microspores just after they have separated by furrowing physically influences the number and location of the three apertures commonly found in dicot pollen. This, of course, then raises the question of why a pollen grain needs more than one aperture for germination.

It should be mentioned here that in some species among various scattered families the four microspores and resulting pollen never separate from each other. Even larger aggregations than tetrads may occur, up to the entire pollen complement of a pollen sac. In such plants, which typically have very reliable animal pollinators or are self-pollinating, packages of 4, 8, 16, 32, or even more pollen are shed instead of single pollen grains. These are not considered here because they are rare among cultivated plants (orchids are the most important exception). Knox (1984) has described in considerable detail the various kinds of pollen aggregates.

DURATION OF MEIOSIS

How long does it take for a mmc to complete meiosis? This is an easy question to ask, but it poses severe technical problems because observation involves killing and processing the very cells being observed; thus one cell cannot be followed through meiosis. By laborious sampling methods, estimates have been obtained for about 70 animal and plant species. Among the 39 species, hybrids, varieties, and cultivars of angiosperms that have been studied, the time range of meiosis is from 18–400 hours, or 0.75 to almost 17 days. Here are examples from cultivated plants, as selected from Bennett (1977):

- Petunia hybrida (18 hours)
- Beta vulgaris (24 hours)
- Pisum sativum (30 hours)
- Hordeum vulgare (39 hours)
- Secale cereale (51 hours)

- Vicia faba (72 hours)
- Allium cepa (96 hours)
- Lilium henrii (170 hours)

Meiosis in the olive tree (Olea europaea, Oleaceae) is also reported to take about 170 hours (Fernandez and Rodriguez-Garcia, 1988). Some other species in the lily family take longer than lily and olive. Why there is such a range of duration times is still fundamentally unknown, but it is known that meiosis can be affected by environmental conditions, most notably temperature; some plants (e.g., walnut) even remain over winter in a stage of prophase. Also, most polyploid species, for unknown reasons, take less time for meiosis than their diploid counterparts. Meiosis is always much slower than mitosis in the same plant, however, mostly because of the lengthy prophase of the first meiotic division.

TAPETAL BEHAVIOR

In the previous chapter the tapetum (Latin for "carpet" or "tapestry") was shown to be typically a single layer of cells lining each pollen sac. Although more than two types of tapetum have been described, two broad categories are recognized by most workers. A tapetum that remains in place as a lining around the sac throughout pollen development is called a "parietal tapetum". A tapetum that flows amoeba-like into the sac interior after the callose dissolves and engulfs the separated microspores is called an "invasive tapetum". These two contrasting tapetal behaviors have been reviewed in great detail elsewhere—e.g., by Pacini (1997). Information about the tapetum presented here is of a more general nature.

The parietal tapetum is also identified by its presumed function as "glandular" or "secretory" tapetum. It is by far the most common type. Present knowledge assigns it to 195 families, of which about 90% are dicots. The less common invasive tapetum is also referred to as "amoeboid" or "plasmodial" tapetum. It is known from only 32 families, of which 21 are monocots (Davis, 1966; Bhandari, 1984). Two important exceptions to these tapetal distribution patterns deserve mention. The grasses (monocot family Poaceae) have been reported to have only a parietal tapetum (Kirpes et al., 1996); an invasive tapetum, uncommon elsewhere among dicots, has been found in all members of the large sunflower family (Asteraceae) studied so far (Pullaiah, 1984; Lersten and Curtis, 1990).

Why two types of tapetum have evolved is unknown. Some have suggested that the invasive tapetum provides contact with all microspores, but this postulated advantage seems unlikely considering the tiny volume of even the largest pollen sacs and the miniscule distances needed for diffusion even from a parietal tapetum.

Parietal tapetal cells typically have two nuclei, but additional nuclei per tapetal cell are also known (Wunderlich, 1954); as an example, common dandelion (Taraxacum officinale, Asteraceae) has up to 16 nuclei per tapetal cell. Multinucleate tapetal cells are a highly unusual kind of plant cell. Also unusual are tapetal cells that have only a single conspicuously enlarged nucleus with endoreplicated DNA, which is DNA replicated without mitosis. In the large legume family the tapetal cells in the subfamily Caesalpinioideae have two to many nuclei, whereas in the subfamilies Mimosoideae and Papilionoideae, which include most cultivated legumes, tapetal cells are reported to always remain uninucleate and often with endoreduplicated DNA (Buss and Lersten, 1975).

Tapetal nuclei are known to replicate DNA to 4–16C levels (Nagl, 1979). Franceschi and Horner (1979) found that the tapetum in starof- Bethlehem (Ornithogalum caudatum, Liliaceae) was a mixture of 64% binucleate cells with about 7C per nucleus, and 36% uninucleate cells with about 13C per nucleus. This is above the 4C level of the mmc's.

Most other investigators have also reported DNA levels above 4C. A study on onion reported that nuclear size and DNA content of tapetal cells increased during meiosis (Castillo, 1988). Only a few tapetal cells became binucleate, however; instead, most nuclei replicated DNA by an irregular internal process termed "cryptic polyteny" and thereby reached DNA levels between 16C and 32C. A mixture of binucleate (78%) and uninucleate (22%) tapetal cells had been reported earlier in tobacco by Scarascia (1953).

These examples of proliferation of tapetal DNA by one mechanism or another suggest strongly that these cells are involved in intense metabolic activity followed by cell death. Similar behavior is known in both endosperm cells and in embryo suspensor cells during early stages of embryo growth.

This unusual tapetal nuclear behavior is conspicuous, but other changes also occur both in the cell wall and in the cytoplasm. The tapetal wall might begin to disappear as early as meiosis, and at some post-meiotic stage it is common for at least its inner tangential wall facing the pollen sac to disappear. Tapetal cells of many plants eject the contents of small membrane- bound vesicles onto this exposed cell membrane, which then form discrete granular bodies called orbicules (or "Ubisch bodies," after their discoverer). The orbicules gradually become coated with sporopollenin, which supports the hypothesis of some workers that the tapetum secretes sporopollenin and contributes to the exine of the pollen wall. Some of this sporopollenin is hypothesized to secondarily polymerize as a by-product around the tapetal orbicules.

Tapetal orbicules are commonly featureless spheroids consisting of a lipid core (Steer, 1977a,b) but in some species they take on the appearance of the patterned exine of the pollen wall. This is especially pronounced in many grasses, where sporopollenin accumulates between the originally separate orbicules to form a continuous tapetal membrane (Banerjee, 1967; Christensen et al., 1972) over the entire inner surface of the tapetum. The tapetal membrane therefore resembles the outer pollen wall in both physical appearance and time of formation. Tapetal orbicules have a possibly practical significance because they occur on the tapetum of some allergenic plants. Vinckier and Smets (2001) illustrated several examples and suggested that these extremely small particles could be important in pollinosis, which is a serious allergenic reaction in the lower part of the lungs.

Tapetal cytoplasm undergoes considerable changes and, surprisingly, in some ways it mimics the mmc's as they progress through meiosis to microspores and pollen. Dickinson (1987) concluded that, In fact, apart from meiosis itself, the sequence of events taking place in these two groups of cells [i.e., tapetum and mmc's] is more or less identical. They undergo precisely parallel phases of synthesis of protein, lipid, and polymers. Indeed, the tapetal cells of some species even attempt to form some kind of patterning on their surfaces [this a reference to the orbicules and orbicular wall mentioned earlier]. In view of the complexity of these events, and of the fact that the tapetal tissue is only one cell in depth, it is indeed remarkable that the distinction between these two tissues is maintained.

Dickinson implies here that the tapetum may be some kind of sterilized sporogenous cell layer. Many insect-pollinated plants eject tapetal substances into the pollen sac that coat the outside of pollen grains. The two recognized types of such tapetal products are pollenkitt and tryphine (Dickinson and Lewis, 1973). Pollenkitt (German for "pollen glue") is typically yellow or orange because it contains carotenoid and flavonoid pigments. It causes pollen grains to stick to each other and to their insect pollinators. Its chief components, however, are various neutral (non-polar) lipids, according to a survey of the pollen of 69 species of 28 families (Dobson, 1988). Tryphines (Latin, meaning approximately "that which is rubbed on") are more heterogeneous but also often rich in lipids. Tryphines are said to help pollen stick to the stigma, but it is often difficult to distinguish between tryphines and pollenkit.

An example of tryphine was shown in the previous chapter from radish and mustard. An example of pollenkitt, occurring as large globules in the tapetal cells of flowering ash, Fraxinus excelsior (Oleaceae), can be seen just before these cells rupture. An appreciation of the amount of pollenkitt that may be produced, and how it causes pollen to clump, is revealed in a scanning electron microscope view into a mature pollen sac of the witch-hazel shrub, Hamamelis mollis (Hamamelidaceae), where all the pollen grains are clumped together in what appears as a syrupy pollenkitt matrix.

TAPETAL FUNCTION

The ephemeral tapetum usually perishes before pollen is mature, but evidence from many studies shows that it has several secretory and nutritive functions related to pollen development, pollination, and even to pollen germination. Whether the tapetum has the same multiple functions in all species is harder to defend. Early investigators recognized that the tapetum was a kind of "...jacket of nourishing cells..." (Coulter and Chamberlain, 1903) that provided unspecified substances necessary for pollen development. In the 1950s it was suggested that chromatin (actual genetic material) moves physically from tapetum to microspores. This proved impossible to defend after a brief period of notoriety. A study by Moss and Heslop- Harrison (1967) on maize anthers, as well as several other studies, failed to detect anything moving from tapetum to developing pollen. What can be said more specifically about tapetal function?

There is direct evidence that pollenkitt and tryphines, substances of tapetal origin, coat the pollen of many species, as shown in the previous section and in the mustard family example described in the previous chapter. Certain proteins in this tapetal exudate may become recognition factors that contribute to a compatible pollen germination response on the stigma. In such species the sporophytic parent, of which the tapetum is a component, therefore indirectly helps the gametophytic pollen grain to gain acceptance by the stigma of the carpel. This tapetal function, however, benefits only mature pollen, its transport, and its subsequent interaction with the stigma.

The degeneration of the tapetum, either by dramatic rupture or gradual collapse, also coincides with the filling up of the vacuolate mid-stage pollen grains with food reserves after their earlier reserves have been depleted (Horner and Lersten, 1971; Christensen and Horner, 1974). Reznickova and Dickinson (1982) reconciled these two functions by providing evidence, at least for lily, that some of the accumulated tapetal lipids are deposited as pollenkitt and some are hydrolyzed and subsequently transferred to vacuolate pollen grains, where they are converted to food reserves.

Stieglitz (1977) produced convincing evidence for a third tapetal function. She reported that a callose-dissolving enzyme, □-1,3 gluconase, occurs in the tapetum of two species and hybrid cultivars of lily, with the highest concentration occurring shortly after the 2nd meiotic division, just before the callose around the tetrads begins to dissolve. She was able to separate the tapetum from the rest of the anther and extract the enzyme from it. She also removed microspore tetrads still encased in callose from pollen sacs and divided them into two sets. One set was treated with the enzyme, and the callose dissolved. The second set went untreated and its callose did not dissolve. This experiment demonstrated that tapetum-produced □-1,3 gluconase dissolves callose (one can wonder as an aside why the tapetum produces this enzyme instead of the microspores themselves).

Another enzyme (phenylalanine ammonialyase) has also been found to be almost restricted to the tapetum in tulip (Tulipa sp., Liliaceae) by Rittscher and Wiermann (1983). They regarded this as evidence for tapetal involvement in phenylpropanoid metabolism in the anther because this biochemical pathway eventually produces the flavonoid pigments that are commonly deposited on the pollen wall as a component of tryphine. This enzyme and its product, a pentahydroxychalcone, were detected by Sutfield et al. (1978) in the tapetum of tulip when callose began to dissolve. The concentration of both substances peaked when microspores began to divide to become vacuolate pollen, after which both enzyme and product dropped rapidly to almost zero by the mature pollen stage.

A recent study of lily using labeled monoclonal antibodies (Aouali et al., 2001) has shown that pectin is produced in the anther wall and is probably secreted by the tapetum into the pollen sac fluid, where it is absorbed into the developing pollen and becomes a significant component of the carbohydrate reserves. They speculate that the pectin later helps to form and maintain the pollen tube wall after germination.

The tapetum has also been postulated to produce the carotenoid precursors that are synthesized into sporopollenin, the chief component of the exine of the pollen wall, or at least that the tapetum makes a substantial contribution to the pool of sporopollenin precursors. Indirect evidence for this is the sporopollenin coating that forms around orbicules on the tapetal cell surface, as shown in the example of sorghum. Not all agree, however. Steer (1977a,b), in two important papers on tapetum development in oats, concluded that this hypothesis is not supported by convincing evidence.

At the level of gene action, studies done mostly on mustard (Brassica) and petunia have identified genes in both tapetum and microspores that code for callase, lipids, flavonols, and oleiosins (storage oils), all of which involve tapetal function (studies cited in Kapoor et al., 2002). These workers have used visualization techniques to show that the gene TAZ1 in pre-meiotic petunia occurs in all tissues except the tapetum and mmc's, but after meiosis it can be detected only in tapetum and microspores. A technique that "silenced" this gene in test plants caused tapetum, microspore, and pollen degeneration. TAZ1 appears to affect flavonol synthesis in the tapetum and/or its transport to the microspores. These molecular studies have provided deeper and more precise insight into certain tapetal functions.

All of the proven and postulated tapetal functions just described cannot be said to operate in all species, but this versatile cell layer is remarkable for its importance in several critical ways to both developing and mature pollen.

POST-MEIOSIS: THE POLLEN WALL

After meiosis the four microspores in a tetrad typically remain together and isolated from other tetrads because the callose persists. It has been suggested that isolation by callose is necessary for the transition to the gametophyte (pollen grain) generation to occur without influence from the surrounding sporophyte (anther tissue). It has also been suggested that callose may act as a template, or mold, that influences the configuration and pattern of the future pollen wall. A detailed study of lily microspores by Dickinson and Sheldon (1986) did not, however, support this hypothesis. Whether or not this proposed function is true, callose does persist until the microspores form at least the rudiments of what will become the pollen wall.

A young microspore, while still encased in callose, initiates the framework for its distinctive pollen wall, which will be completed much later. The mature pollen wall consists of two distinct and chemically different layers. The outermost layer is the exine, which is composed largely of a durable material called sporopollenin, a polymer that can remain extremely resistant to physical and chemical agents for millions of years under anaerobic conditions. This is why pollen deposited in oxygen-poor strata—e.g., peat bogs and lake beds—can be dug up and studied as exine "skeletons" after eons. This property of inertness has also defied attempts to adequately analyze the chemical composition of sporopollenin.

The inner and far less–resistant wall layer is the intine, which is composed largely of carbohydrates, including cellulose, and is similar in some respects to the primary cell wall of a vegetative cell (Nepi and Franchi, 2000). Exine and intine can each be subdivided into two or more subsidiary layers, but not everyone agrees on how many layers or what to name them, especially because such structural details differ considerably among species.

Pollen wall development has been studied extensively and intensively by light microscopy, electron microscopy, and by various other techniques down to the molecular level. The details of these topics lie mostly outside this book. Instead, a series of observations on sorghum by Christensen et al. (1972) is presented, which is not only representative of cereal grasses but also illustrates general features of pollen wall development. Entry into the great number of studies that provide more detailed descriptions of the pollen wall can be found in Dickinson and Sheldon (1986), Knox (1984), Rowley (1981), and Scott (1994).

At this early tetrad stage the microspores have just a naked cell membrane. The first visible event is the appearance of some vaguely fibrous material called primexine, which provides a temporary matrix within which the permanent exine will develop. Denser material can be detected within the primexine as callose begins to dissolve from around the microspore.

This new material appears slightly thicker and more irregular just after the callose dissolves completely; by the early vacuolate microspore stage it acquires the faintly recognizable form of a median dense layer studded with thicker deposits at intervals and separated from the plasmalemma by a thin basal layer of primexine. During the vacuolate microspore stage the exine continues to thicken and take shape, and gradually its component zones become distinct. The microspore enlarges, often to several times its original volume, while the exine develops, so the exine must retain flexibility as it seemingly becomes thicker and less flexible. A certain degree of exine flexibility is maintained even after pollen is mature, as shown experimentally by Rowley and Skvarla (2000) and as is also demonstrated by observations of pollen deformation that occurs during harmomegathy.

But exine formation is just the first stage of pollen wall development, even though it appears to be the most complex part. The intine is deposited later between the cell membrane and exine after the first microspore mitosis creates the vacuolate pollen grain. The intine attains maximum thickness rather late in development while the pollen is becoming engorged with food reserves and using some of them for intine development. In many species the completed intine often equals or exceeds the exine in thickness. The intine is composed chiefly of carbohydrates but it is not homogeneous. Three distinct layers have been identified in some plants: a thin outer layer with pectic microfibrils, a main middle layer with embedded inclusions of proteins and enzymes, and an inner layer with callosic and pectic components. This inner zone will expand and become the pollen tube wall upon germination (Heslop-Harrison, 1987).

The intine is often noticeably irregular in thickness and permeated by cytoplasmic channels. The vegetative cell inserts enzymes and proteins into these channels and into isolated pockets of vegetative cell cytoplasm embedded elsewhere in the middle layer of the intine; these tiny but critical substances from the pollen grain will be released onto the stigma as gametophytic recognition factors, which are required for successful pollen germination of many species.

The magnification is too low to show microchannels, but the intine is obviously thicker near to, and beneath, the pore. Microchannels are reported to traverse the exine in many species (Rowley et al., 1987), and they could transport substances to the pollen surface or help adapt pollen to harmomegathy. A summary of wall formation in the sorghum pollen grain correlated with events of orbicular wall formation on the tapetum includes labels for a few of the common exine subdivisions (tectum, cavea, nexine). Pollen wall features vary greatly among angiosperms, and different names have been applied by different workers, but sorghum shows a representative sequence of pollen wall development.

Most pollen grains have one or more circular- to-elongate sites called apertures. In these local areas the exine either does not form, which exposes the intine, or it is very thin. A pollen tube will emerge through an aperture, even though the intine is typically thicker around and below most apertures than elsewhere, as for example in sorghum. In apertures of some pollen, the intine even bulges outward to form what is called an oncus (Greek for "lump"). The thickened intine at the site of an aperture possibly indicates that more recognition proteins and enzymes occur here. Apertures vary in size, shape, and number and range from elongate (colpi) to circular (pores).

Pores of grass pollen and monoaperturate pollen of some other monocots also have a distinctive circular island of exine, called an operculum (Latin for "lid"), which fits within the pore much like a loose cork in the mouth of a bottle. A combination of pore and colpus, called colporate, is common among dicots. Most monocots have either one colpus or pore, but there are exceptions. Some plants even have a uniformly thin exine with no apertures (inaperturate), so a pollen tube presumably can emerge anywhere. Speculations about why apertures vary so much in shape and number involve both physical and functional arguments, but none are widely accepted as yet. Pollen wall architecture in general is seemingly endlessly varied, and detailed description and classification of wall features, as well as attempts at explanation, occupies a great part of the specialized discipline of palynology.

Pollen of cultivated plants exhibit most of the possible aperture arrangements; here are some examples of aperture types from among them: Inaperturate (no aperture): cottonwood, Populus deltoides (Salicaceae) canna lily, Canna indica (Cannaceae) Monoaperturate (one colpus or pore):

grasses

Triaperturate (three colpi or pores):

legumes (Fabaceae)

mustards (Brassicaceae)

roses (Rosaceae)

carrot (Apiaceae)

Polyaperturate (many colpi or pores):

beet (Chenopodiaceae)

cucumber (Cucurbitaceae)

poppy (Papaveraceae)

Tryphine and pollenkitt, two tapetal products mentioned earlier in this chapter, are not part of the pollen wall; but in many pollen grains, particularly those of animal-pollinated species, these external deposits do seem to function like an extra wall layer. It has been suggested that these coatings seal the pollen grain to reduce water loss in addition to adhering pollen grains in clusters to aid pollination and release of sporophytic proteinaceous factors important for interaction with the stigma.

POST-MEIOSIS: INTERNAL MICROSPORE/POLLEN EVENTS

After a microspore enlarges in volume to the size determined for its species, it divides mitotically to form the generative and vegetative cell of the pollen grain. This may occur before the microspore becomes conspicuously vacuolated, in other species it may occur after a large central vacuole has formed. A parabolic (curved) cell plate commonly forms during telophase, which after cytokinesis results in a small lens-shaped--to–spheroidal generative cell pressed against the vegetative cell membrane. This unequal partitioning of cytoplasm occurs in all pollen studied.

Experiments inducing dislocation of microspore cytokinesis have produced two equal-sized pollen cells, both of which act like vegetative cells and are nonfunctional (Twell, 1994). Scheres and Benfrey (1999) felt that because plant cells are constricted within a wall they require an asymmetrical division to induce differentiation rather than simply duplication of the same kind of cell. But they admitted that there is no explanation as yet why this should be necessary.

The newly formed generative cell is separated from the vegetative cell for a short time by a thin layer of callose that lies along the former parabolic cell plate. Górska-Brylass (1970) observed 82 species of dicots and monocots and found that the callose persists for about 12 hours, and then disappears while the generative cell is still appressed to the pollen wall. Górska-Brylass thought it was significant that the generative cell is isolated by callose during the period just preceding its own DNA synthesis. Such physical isolation has also been suggested to explain why callose isolates the mmc's and the megaspore mother cell in an ovule. This callose seems to have no relationship to the subsequent migration of the generative cell into the interior of the pollen grain.

The generative cell moves away from the wall and into the interior of the vegetative cell after callose dissolves, which is perhaps a unique example of one plant cell becoming completely surrounded by another cell. A curious exception occurs in *Rhododendron* (dicot family Ericaceae), where the generative cell has one long cytoplasmic tail that attaches it to a polysaccharide knob on the inside of the pollen wall and a second looping tail that ends blindly in the vegetative cell cytoplasm (Theu nis et al., 1985). The function of this novel arrangement is unknown, but it may prevent the generative cell from entering the pollen tube before the vegetative nucleus, if such a configuration is required.

Generative cells typically become ovate to variously elongate while in the pollen grain. They often lack plastids, possibly because before microspore mitosis the plastids usually migrate to an area of the vegetative cell away from where the future generative cell will form, as shown, for example, by Sanger and Jackson (1971) in the blood lily. Exclusion of plastids from the generative cell has also been shown by transmission electron microscopy in two species of garden balsam, *Impatiens* (dicot family Balsaminaceae) by van Went (1984), and from generative cells of petunia, apple, cherry, tomato, tobacco, and olive by Cresti et al. (1984).

More varied plastid behavior was observed among several species by Hagemann (1981), but none ever entered the egg cell. All plastids were excluded from generative cells of 11 species, including barley, beet, cotton, and tomato, but in a few species—e.g., potato (*Solanum tuberosum*)—some plastids were included but degenerated later. In other cultivated species—e.g., flax and rye—several plastids remained intact in the generative cell, were incorporated later into the sperm cells, but were jettisoned into the embryo sac when the sperm nucleus entered the egg cell. Plastid distribution in generative cells of several members of the monocot family Liliaceae was presented in a series of six nicely illustrated papers, and Miyamura et al. (1987) followed plastid distribution in the generative cell and/or sperm cells of 16 species of dicots and monocots using fluorescence light microscopy and special stains to detect DNA-containing "nucleoids" of these organelles. All of these studies described varied plastid behavior among species: initial exclusion, later degeneration, or persistence of plastids until some unknown later time, because they did not follow the pollen tubes until fertilization occurred.

A more ambitious survey by Corriveau and Coleman (1988) used fluorescence microscopy to detect plastid DNA in either the generative cell or the sperm cells, from pollen of 235 species representing 80 families. No plastids occurred in 192 species, but plastids were found in the other 43 species. They concluded, therefore, that plastids in generative cell or sperm cells are rather uncommon. It is of interest that among agronomically important families they found no plastids in 18 grass species, which includes the common cereal grasses, and none in any of the 7 crucifers (Brassicaceae) tested. Among 39 legumes studied, however, 9 species had plastids, all restricted to the Phaseoleae (bean tribe), Trifolieae (clover tribe), and Vicieae (pea tribe).

These studies on generative and sperm cell plastids were stimulated by earlier genetic studies that concluded plastids can sometimes be inherited from the male, and sometimes not, findings of great interest regarding the possibility of extranuclear inheritance. Plastid inheritance via sperm cytoplasm is not ruled out completely now, but it seems improbable for most species in light of these studies.

Extranuclear inheritance of mitochondria is more difficult to trace. They are thought to be discarded by the sperm cells upon entering the egg cell, but evidence is still uncertain. DNA of the generative cell nucleus stains intensely, and its chromosomes are usually reported to remain in a condition that resembles prophase (Steffen, 1963). The vegetative cell nucleus, in contrast, stains faintly, appears to be quiescent, and will not divide again. The generative cell will divide mitotically once more to form the two sperm cells, often before the pollen grain becomes completely engorged; barley is a typical example. The tiny sperm cells each have a nucleus and a small amount of cytoplasm. They are often described as lacking a cell wall, but a very thin wall exists. Some sperm cells are greatly elongated, and in many species they are physically connected to each other and to the vegetative nucleus, forming a male germ unit.

Pollen of most species is shed from the anther with just a generative and a vegetative cell (bi-celled), but in some species the generative cell has already divided to form two sperm cells (tri-celled) before the pollen is shed. An important survey by Brewbaker (1967) showed that 30% of the species in his sample of 2,000 dicots and monocots were tri-celled. Tri-celled pollen was scattered among many families, which suggested that it has evolved independently many times. Most economic plants have bi-celled pollen, but important exceptions are the grass family (Poaceae) and sunflower fam- ily (Asteraceae), in which all pollen appears to be tri-celled. There have been few studies comparing bi-celled with tri-celled pollen, but those conducted so far indicate differences in several physiological characteristics.

The vegetative cell has virtually all of the starch or lipid food reserves and occupies most of the pollen grain. Most information about pollen reserves is from the Baker and Baker (1979) survey of mature pollen of almost 1,000 species representing 124 families of dicots and monocots. They measured pollen diameters and used simple chemical tests to determine whether pollen was starchy (mostly starch but with some lipid) or starchless (mostly lipid but with some starch). No pollen had either starch or lipid exclusively as a food reserve. They found that starchy pollen had a greater size range (mean diameter of 62.6 ¼m) than did starchless pollen (mean diameter of 36.7 ¼m). Grasses, for example, have large starchy pollen.

DURATION OF POLLEN DEVELOPMENT

The examples given earlier in this chapter for the duration of meiosis are a very small sample. The time required to complete all of pollen development is also sparsely documented. The annual tomato species, *Lycopersicon peruvianum* (Solanaceae), takes only 13 days to develop from mmc's to mature pollen (Pacini and Sarfatti, 1978). Wheat is also an annual and its pollen develops almost as quickly as that of tomato, requiring 15–16 days (Xiang-Yuan and De-Cai, 1983). Pollen of the sweet cherry tree, *Prunus avium* (Rosaceae), requires approximately 39 days to develop (Pacini et al., 1986), which is about three times longer than tomato pollen. Pollen of the olive tree, *Olea europaea* (Oleaceae) is even slower, taking 44 days to develop (Fernandez and Rodriquez-Garcia, 1988). The olive time period is subdivided into 7 days for meiosis, 3 days for the tetrad in callose, a further 7 days for the released microspore stage, and a final long period of 27 days from microspore mitosis to the mature engorged pollen. These few examples reveal a range of about 2–7 weeks from mmc to mature pollen grain. One could speculate that pollen develops more slowly in woody plants than in herbs, but many more studies are needed to establish any trends.

GENE EXPRESSION DURING POLLEN DEVELOPMENT

The review by Twell (1994) included detailed information about the expression of genes in microspores and pollen. The evidence reveals that there are 20,000 to 24,000 unique messenger RNAs (mRNAs) in the mature male gametophyte. This measure of gene number is considerably less than the 30,000 unique mRNAs in the sporophyte, which is not surprising since the pollen grain is much reduced in size and has a narrower range of activity. Different sets of genes are operational in microspores, mature pollen, and when pollen is interacting with the stigma prior to, and during, germination on the stigma.

NUMBERS OF POLLEN PRODUCED

The question of how many pollen grains are produced must be qualified by whether reference is to pollen sac, anther, flower, or an entire plant. Few estimates can be found for any of these but some absolute and indirect figures are available for some plants—e.g., the number of grains produced per ovule by a flower. It is of course impractical to count all pollen grains except for a few unusual species, and so statistical sampling is required, which yields estimates that can vary considerably from actual numbers.

De Vries (1971) found different numbers of grains per anther among wheat varieties, but the collective range was 1,750–2,850. He cited two other wheat studies that reported a collective range of 2,000–8,000 grains per anther. Considering that a wheat flower has three anthers but only one ovule, this means that each flower produces somewhere between 5,250 and 24,000 pollen grains to ensure that just one of them will fertilize a single ovule. For comparison, a maize flower produces over 10,000 grains, and a rye flower manages to manufacture close to 57,000 grains. All three of these grasses depend on inefficient wind-pollination. Closer to the other extreme is lettuce; one can estimate from the illustrations in Jones (1927) that only about 250 pollen grains form in each anther of this largely self-pollinated flower. With five anthers and only one ovule, perhaps only about 1,300 grains are produced by a lettuce flower to ensure fertilization.

The examples just mentioned include numbers of pollen produced per ovule as well as the total number per flower, because there is only one ovule per flower. But since ovule number (and therefore seed number) per flower varies greatly among plants, a more accurate comparison among species can be made by dividing the number of pollen grains by the number of ovules, which gives the pollen/ovule ratio for a flower.

A survey by Cruden (1977) used this ratio and revealed an enormous range among many dicots and monocots. He counted few pollen grains per ovule (only 3–7) in cleistogamous (permanently closed) flowers; more but still few (18–39) in flowers that are obligate selfpollinators; followed by facultative self-pollinators, which are mostly selfing but sometimes outcrossing (32–397); facultative outcrossers, which mostly outcross but sometimes self-pollinate (160–2,558); and obligate outcrossers (1,062–19,525—this category, of course, includes species with separate staminate and carpellate flowers on different plants).

The trend is clear: the riskier the chance of successful pollen transfer, the more pollen is produced per ovule.

Consider the wind-pollinated giant ragweed (*Ambrosia trifida*, dicot family Asteraceae). One pollen sac of one anther contains thousands of pollen grains. Multiply that by four pollen sacs per anther, then again by hundreds of male flowers per plant as compared to only 2–3 female flowers per plant, each with only one ovule.

The total number of pollen per plant is unknown but certainly almost of astronomical proportions, which makes it obvious why the late summer air in many regions is thick with ragweed pollen.

A study of pollen/ovule ratios and correlated characteristics in the economically important legume tribe Trifolieae (Small, 1988) showed a range extending from only 312 pollen grains per ovule in *Medicago* (alfalfa genus) to 2,618 in *Melilotus* (sweet clover genus). Small found that the smaller ratios were correlated with larger pollen grains—e.g., the mean pollen volume in *Medicago* is over three times that of pollen in *Trigonella* and *Melilotus*.

Other factors add complications, such as annual versus perennial species, number of self-incompatible flowers open at any one time in the same inflorescence, and pollination strategy. Thus the pollen/ovule ratio needs to be "fine-tuned" when used within a restricted group of plants. Cruden (2000) recently reviewed the subject and provided a re-synthesis of his pollen/ ovule hypothesis to explain its influence on pollen production

5

Carpel and Gynoecium

The gynoecium and the carpels that compose it are complex entities that require an understanding of the technical terms that describe them. That effort will also reveal some fundamental similarities to the stamen and the pollen it produces. So it will be helpful here to first review briefly the terminology of stamen and androecium. All stamens in a flower collectively comprise the androecium ("male household"). Each stamen is a microsporophyll with an anther that contains four microsporangia (pollen sacs). Each microsporangium produces many microspore mother cells; these undergo meiosis, each producing four microspores. A microspore germinates internally and becomes a two- or three-celled microgametophyte (pollen grain). Five of these formal terms have the prefix "micro" in common, which can be thought of as indicating "maleness" as well as small size.

There is a parallel set of technical terms and concepts for the female side. The stamen counterpart here is the carpel (Greek, meaning "fruit"). All carpels of one flower collectively constitute the gynoecium ("female household"). The formal name for a carpel is megasporophyll, which means "leaf that bears megaspores." The megasporophyll first produces ovules, each of which could be likened to a megasporangium. Each ovule produces only one megaspore mother cell; it undergoes meiosis to produce four megaspores, but only one of them will survive. The megaspore germinates internally, and after usually three mitotic divisions it forms a 7-celled megagametophyte (embryo sac). Up to this stage one can appreciate that there are similarities with pollen development and also with the terminology, only substituting "mega" (representing "femaleness" and "large") for "micro."

The megagametophyte, unlike the pollen grain, is not released but remains within the ovule, which in turn occurs within a carpel.

CARPEL EVOLUTION AND DEVELOPMENT

A stamen withers after its pollen departs, but a carpel persists because it retains its female gametophytes. After fertilization the gynoecium— of which the carpel is a part— matures (often with considerable structural changes) into a fruit that encloses the seeds. All flowering plants develop their seeds within carpels, a characteristic structural feature that provides the formal name for the group: Angiospermae (Greek for "covered seed").

Looking at most modern flowers will not convince anyone that a carpel has anything in common with a leaf. Furthermore, post-fertilization physical changes can make it an almost unrecognizable version of the pre-fertilization carpel. But in spite of this seemingly unlikely comparison, there is good evidence that the earliest carpel was indeed a simple leaflike structure studded with two rows of ovules, one row parallel to each margin. This originally open ovule-bearing leaf, descriptively speaking, at some point in evolutionary time folded or rolled itself longitudinally so that the margins became appressed to each other, thereby producing a true closed carpel. Among a few modern tropical families one can find what is essentially a folded-leaf carpel lacking stigma and style.

Pollen can land and germinate anywhere along the extended suture where the carpel margins join and send pollen tubes directly inward to fertilize nearby ovules. In most present-day angiosperms, however, carpels have evolved into more elaborate structures. According to a widely accepted hypothesis, the first step was the elimination of the uppermost ovules in a folded-leaf carpel, followed by the transformation of this now-sterile apical part of the carpel into a tube or column (Greek word is "style") with a specialized tip called a stigma (Latin for "mark" or "scar"— probably because it often withers after pollination, or if pollination does not occur). Each of these new carpel structures (stigma, style, ovary) gradually acquired anatomical and physiological features to protect the ovules, control access to them, or both.

Why did the original folded leaf type of carpel, which seems on first thought to be a simple and sensible arrangement, become modified to stigma, style, and ovary? There are several probable advantages:

- It allows ovules and resulting seeds to be enclosed more securely for protection from predators.
- An intervening style removes the ovary and its ovules some distance from potentially predatory pollinators landing on the stigma.
- The style can influence (stimulate or inhibit) pollen tube growth chemically, nutritionally, or even by anatomical means.
- A stigma concentrates pollen grains in a small area, where they can be accepted or rejected by positive or negative chemical interaction.
- Carpel ovaries can adhere to each other or merge in various ways and still retain a style and stigma where pollen can land.

There are probably other reasons why almost all carpels have evolved beyond a simple folded leaf configuration, given the spread of flowering plants to virtually all climatic zones and most terrestrial and watery habitats, and their adaptation to various pollination mechanisms. A single carpel first appears as a tiny mound (primordium), which then becomes more-orless cuplike or pitcher-shaped because a ridge grows up around its margin. This ridge grows asymmetrically, with three sides typically becoming exaggeratedly elongate and little or no growth occurring on the remaining side. Such asymmetry produces an elongate vertical slit. One can appreciate that the folded-leaf configuration in a modern simple carpel therefore does not result from a leaflike structure that forms first and later folds on itself. He also (1981, 1983) reviewed previous studies of carpel development and made scanning electron microscope observations on numerous additional species.

CARPEL VARIATIONS: GENERAL CONSIDERATIONS

Carpel number, shape, and degree of merger with other carpels has evolved in many different ways, partly because carpels do not first become leaflike, which would severely limit any further developmental changes. Such changes are initiated instead at or near the primordial stage, when the incipient carpel or carpels are most plastic and capable of radical shifts in development. This has allowed carpels and gynoecia to adapt their structure and physiology to various kinds of pollinators, to develop features that allow only compatible pollen to germinate, and to evolve tissues and structures to guide and nurture the resulting pollen tubes. There are, however, remarkable developmental changes that often occur in the carpel wall after fertilization, which has allowed the gynoecium to evolve into the fruit in myriad ways to provide protection for seeds and dispersal of the seeds by animals or by physical means.

Grasses have what seems at first to be a simple carpel with one ovule, but detailed examination reveals a highly evolved gynoecium composed of three fused carpels, of which two have disappeared except for evidence of them in the form of a pair of plumose styles and stigmas. The development of a grass gynoecium shows some resemblance to the pattern of development found among true single carpels just described, but it is worthwhile showing an example for comparison.

In maize, as shown beautifully in a scanning electron microscope study by Cheng et al. (1983), there is first a mound, followed by development of a gynoecial ridge that gradually surrounds the ovule primordium. As the gynoecial ridge begins to grow over the ovule, two bumps form on one side of it, marking the initiation of the styles. The gynoecial ridge gradually overtops the ovule except for a small cleft that marks the stylar canal; the two stigma silks continue to elongate.

Just as most flowers have an androecium with more than one stamen, so most flowers have a gynoecium with more than one carpel. Then what about the still commonly used term "pistil"? Pistil is a variant form of the word "pestle," the hand-held grinding rod of a mortar and pestle. The gynoecium of many plants does appear pestlelike, and the 18th century botanist who coined this term simply made use of this familiar tool. The only problem with "pistil" is that it can refer both to a single carpel and to the entire gynoecium; thus it is not precise enough for careful use and is now generally avoided. Another related problem of terms has not been adequately solved: the ovule-bearing part of a single carpel is called the ovary, as is the ovule-bearing part of a variously merged set of carpels. This problem seems not to have caused much, if any, complaint and it will therefore be ignored here.

CARPEL VARIATIONS: APOCARPY

A single carpel per flower occurs in only 11% of all angiosperms. Legumes have the most familiar examples, and since this family has perhaps 18,000 species, it constitutes most of this category. The familiar simple pea or bean pod is therefore a rather uncommon type among flowering plants. The almond flower (*Prunus amygdalus*) and its relatives in the Rosaceae also have a simple gynoecium consisting of one carpel with a clearly distinguishable stigma, style, and ovary—the latter with just two ovules. Even less common than a single carpel are several separate carpels per flower, an arrangement called apocarpy, which occurs in only about 6% of species. Included here, however, are such familiar members of the Rosaceae as strawberry and blackberry.

CARPEL VARIATIONS: SYNCARPY

The vast majority of angiosperms (about 83%) not only have more than one carpel per flower, but the carpels either adhere to each other or actually merge to various degrees, which results in syncarpy ("joined carpels"). The manner and degree of carpel adhesion or merger takes many forms among, and even within, families, as Endress (1982) has pointed out in a review. In a barely syncarpous flower the carpel ovaries merely adhere superficially and each carpel retains its own style and stigma.

Such carpels seem superficially to be syncarpous but each is really functionally independent, interacting with pollen as if it were apocarpous. At the other extreme of syncarpy is the complete integration among all parts of all carpels to form a gynoecium that looks from the outside like a single carpel, with one style and stigma (e.g., the *Citrus* flower).

The syncarpous grass gynoecium mentioned earlier is almost as completely integrated. Between these extremes are all possible intermediate variations. In the apple flower, for example, the five individual carpels that comprise the gynoecial ovary are fused, but there are five separate styles and stigmas. Members of the mustard family (Brassicaceae) have two carpels that approach com- plete merger except for a thin longitudinal septum that divides the gynoecial ovary into two carpel ovaries, as in radish.

An understanding of the possible ways in which carpel ovaries can fuse, both developmentally and in an evolutionary sense, can help one interpret the otherwise often confusing patterns of ovule distribution and individual carpel delimitations within the gynoecial ovary. Keep in mind that in real plants the process of fusion starts at a primordial stage, not after each carpel first develops separately. The dorsal bundle can be thought of as corresponding to the midvein of a leaf. These three major bundles are usually retained even when carpels merge. The two ventral bundles supply nutrients to the ovules and usually end in the upper part of the ovary, but the dorsal bundle continues upward, traversing the style and often extending into the stigma.

If the three carpels each close individually and just adhere to the others, a syncarpous ovary such as that could result. The structural features of each individual carpel are retained, and each would function independently, but the independent closure of each carpel converges the ovules in the common central axis, an arrangement called "axile placentation." Placentation refers to the arrangement of ovules within the ovary, the term placenta (Greek, meaning "a flat place," a term borrowed from zoology) denoting where ovules are attached to the ovary wall. Each segment of an orange or grapefruit is a carpel ovary, and if you cut a citrus fruit transversely you will find that the seeds are attached centrally to the fibrous axile placenta. Returning now to the original three free carpels, imagine that their margins fused before they closed individually, an arrangement like three people holding hands to form a circle. The ovules would now be arranged around the periphery, which is called "parietal placentation." Willows (*Salix*) and poplars (*Populus*) (both dicot family Salicaceae) are examples.

The three open carpels might fuse with each other as they are closing, resulting in a true syncarpous ovary. The septa of each carpel fuses with its neighbor, but with axile placentation and close association of the ventral vascular bundles with each other. Further evolution from this type of gynoecium could involve the loss of all but the central axis of the combined septa, leaving the ovules attached to a free-standing central pole extending up the middle of the gynoecial ovary. This is where free-axile or free-central placentation occurs, for example, in the carnation, *Dianthus caryophyllus* (dicot family Caryophyllaceae). The two ventral vascular bundles of each carpel ovary remain within the central pole, which is therefore a composite septal remnant contributed to by each of the three carpels.

This central post may in turn become shortened, in some plants to a mere stub rising from the base of the gynoecial ovary. The number of ovules are therefore also reduced, often to just one, and they (or it) would appear to arise from the base of the carpel ovary, a condition called "basal placentation". Buckwheat, spinach, beet, and many other plants have a single basal ovule. Remember, however, that there is more than one evolutionary pathway leading to a single ovule per ovary.

The syncarpous modifications of carpels just described can be obscured in one or more ways: a greatly expanded placenta, which can accommodate many ovules (tomato is a good example of this); an increase in the number of carpels comprising the gynoecial ovary; extreme reduction in one or more parts of the carpel; or changes in size, shape, or relations to other flower parts of the existing carpels.

CARPEL VARIATIONS: RELATIONS TO OTHER FLOWER PARTS

Complications do not end with those just described. The relationship of the gynoecium to other parts of the flower can also differ between or within families. If the gynoecial ovary remains free from other floral parts and appears to arise from the center of the flower (i.e., closest to the apex, which satisfies the criterion that it is the last type of appendage to form), it is described as either a "superior" ovary (i.e., gynoecium is attached above all other parts), or the flower is called "hypogynous" (i.e., other flower parts are attached below the gynoecium). A citrus flower is a good example. In contrast, the gynoecial ovary in some plants appears to be embedded in the underlying receptacle of the flower. Such flowers have an"inferior" ovary (i.e., it appears to be attached below other flower parts), or the flower is described as "epigynous" (other flower parts appear to be attached above the gynoecium), as in the apple flower.

Flowers in some families are intermediate because the gynoecium is not sunken but the basal part of the sepals, petals, and perhaps even stamens, is either fused around the lower half of the ovary or they just adhere to it superficially. The ovary therefore appears to be only partly embedded, a condition called either a half-inferior ovary or semi-epigyny. If the cuplike fusion product of these floral appendages extends up around the ovary but does not become fused with it, this cup is called a hypanthium, as in the almond. Only a general and simplified description of the range of possible variations of the gynoecium has been described here. But knowing the common variants of carpel number and their fusion, and other possible mergers with the receptacle and basal parts of other floral appendages, can help in interpreting the flower structure of economic plants.

CARPEL STRUCTURE: STIGMA

The stigma is usually distinct from the style, often appearing as a knoblike swelling. It is surprisingly sophisticated both anatomically and physiologically, however, for what one might think is the simple task of merely capturing pollen grains. The stigma is a "gatekeeper" that can stimulate germination of compatible pollen or inhibit, or at least fail to stimulate, germination of foreign or otherwise incompatible pollen. The functional aspects of the stigma will be taken up in some detail in Chapter 6; here it is appropriate to just mention the structural variations found among stigmas in general.

The stigma is commonly thought of as a knob covered with a viscous fluid. Many plants do have such a wet stigma, as in wild lemon, *Citrus limon*, but it should be emphasized that stigmas of most plants lack any obvious surface secretion, and are therefore called dry stigmas. This is an important distinction among stigmas, as Heslop-Harrison and Shivanna (1977) concluded from their broad survey of about 1,000 species of living plants representing about two-thirds of angiosperm families.

Wet and dry stigmas occur in virtually identical proportions both in dicots and in monocots, indicating that stigmas have evolved along similar pathways in both groups. This distribution suggested to Heslop-Harrison (1984) that the two types of stigma are " . . . more likely to be of physiological than of phylogenetical significance."

Wet stigmas of many dicot and monocot species have a smooth epidermis, but others have papillose epidermal cells. The copious viscous liquid that covers the stigma when it is receptive to pollen may be secreted by epidermal cells of the stigma, or in some species it is first secreted internally from cells that line the hollow style and forced outward from there onto the stigma surface.

The secretory product in still other species is produced in subepidermal pockets within the stigma and later exudes onto the surface, as in some members of the Solanaceae (Konar and

Linskens, 1966). In sweet cherry, *Prunus avium* (Rosaceae), and in several other species, extensive lacunae form within the stigma, which may contribute secretions to the surface as well as aid gas exchange during early pollen tube growth (Uwate and Lin, 1981).

Dry stigmas are about twice as common as wet stigmas, according to Heslop-Harrison and Shivanna (1977). In a later contribution, Heslop- Harrison (1984) emphasized that a modest secretion does occur in some dry stigmas, but that a thin film forms over it, a membrane-like pellicle (Latin for "thin skin") consisting of glycoproteins and lipids that is secreted through microscopic gaps in the underlying cuticle.

One of the glycoproteins has been identified as the binding lectin concanavalin A, a molecule that might help compatible pollen grains adhere to the dry stigma. Epidermal cells of dry stigmas may be either papillose or smooth, just as in wet stigmas. A unique type of dry stigma is the plumose, or feathery, stigma with receptive cells dispersed in multiseriate branches, which is found only in the grass family (Poaceae).

Soybean and many other legumes exhibit certain features of both wet and dry stigmas (Lord and Heslop-Harrison, 1984; Tilton et al., 1984).

There is considerable secretion above and between the long, papillose epidermal cells, but it is covered by a tough pellicle. The pellicle is eventually disrupted by internal pressure from continued internal secretion or from a pollinator that lands on it, which exposes the stigmatic secretion.

CARPEL STRUCTURE: STYLE AND TRANSMITTING TISSUE

The style lies between the stigma and the ovary. It may be solid or hollow, long and slender, short and thick, or so short as to be virtually absent. Can a single common and necessary function be assigned to the style in view of this considerable range of variation? Why is a style necessary? There is at present no explanation as to why a style is almost or completely absent in some plants, but there has been speculation about why various longer styles occur. Some possibilities for the existence of a style include the following:

- To provide a spatial separation of stigma from the ovary to lessen the chances of ovules being eaten or accidentally damaged by pollinators
- To act as a zone of intimate cell-to-cell contact where incompatible pollen tubes can be slowed or stopped before they reach the ovary
- To provide nutrition for compatible pollen tubes but not incompatible ones, thereby causing the latter to slow down and lose the race to the ovules
- To act as a bottleneck—i.e., have an anatomical feature that allows passage to only the first and fastest pollen tubes Evidence for the last function has been shown, for example, in maize. As is true for other aspects of the flower where variation is seen, the factors that determine style length must be studied mostly in individual species.

A theoretical study by Lankinen and Skogsmyr (2001) concluded from computer modeling that "pistil length" (presumably they meant style length) has an effect on selecting pollen tubes by their growth rate (better quality pollen grows fastest), which means that pistil morphology can indirectly select for pollen quality. Cruden and Lyon (1985) addressed the question of whether a pollen grain must be large enough to provide sufficient nutrients to sustain a compatible pollen tube for the entire length of the style; in other words, does successful fertilization depend on matching the size of a compatible pollen grain to the length of the style?

Their measurements revealed a positive correlation only between pollen size and distance to the start of transmitting tissue in or below the stigma. Their results therefore support the hypothesis that the style nurtures pollen tubes, which means that pollen size does not have to be correlated with style length. The style of an individual carpel usually has only the dorsal bundle as its major vein, which continues upward from the carpel ovary. Ventral bundles that supply the ovules usually stop at the apex of the ovary. Also traversing the style is the so-called transmitting tissue (also called stigmatoid tissue), which is either a strand of elongate, thick-walled cells (closed style) or an interior tube, usually lined with secretory cells (open style). Both closed and open types of transmitting tissue have two certain functions:

- To provide physical guidance for pollen tubes
- To provide nutrients for pollen tubes

A third possible function is the secretion of substances that either stimulate or inhibit pollen tubes. Transmitting tissue may continue in a modified form down into the ovary and direct pollen tubes close to, if not directly into, individual ovules. But often there is no guiding tissue or pollen track in the ovary, so that pollen tubes grow more-or-less haphazardly around the ovary wall until they contact ovules by chance.

Transmitting tissue in a closed style begins in the stigma, often just below the epidermis, and continues as a strand of elongate, living cells down the length of the style. Transmitting tissue cells have either a thick, pectin-rich wall or they secrete some kind of extracellular nutritional substance. Pollen tubes grow within or along the wall, or through the extracellular material, but they do not penetrate any transmitting cell protoplasts. Closed styles are more common in dicots than in monocots, although many exceptions occur. And as might be expected, intermediate tissue arrangements also occur in some species.

An example of a closed style is seen in the apple gynoecium, and a magnified cross-sectional view shows that transmitting tissue occupies a considerable volume. A still closer view shows the solid-appearing dark intercellular material secreted by the living transmitting cells, which is probably a gelatinous substance rich in proteins and polysaccharides. In sweet cherry, also a member of Rosaceae, the transmitting tissue has been shown to begin just below the papillose epidermis of the stigma; thus cherry pollen tubes need only grow between the stigma papillae and they immediately contact transmitting tissue. Indirect evidence for its nurturing function in sweet cherry is the presence of starch in transmitting tissue cells and its absence from other cells of the style and stigma (Stosser and Neubeller, 1980).

Transmitting tissue in tomato resembles that of apple, but the tissue is arranged somewhat differently. Kadej et al. (1985) showed that it begins as a broad column in the stigma, gradually narrows in the style, and about halfway to the ovary it splits into two narrower strands of transmitting tissue, each of which enters one of the carpel ovaries. A similar arrangement of transmitting tissue occurs in tobacco, another member of the Solanaceae (Bell and Hicks, 1976). The transmitting tissue cells in both of these examples are alive, with thick walls through which the pollen tubes grow.

An example of a closed style in a herbaceous dicot, wild radish (*Raphanus raphanistrum*, family Brassicaceae), was described by Hill and Lord (1987). The radish ovary is divided longitudinally by a thin septum; thus there are two carpels, with parietal placentation. Transmitting tissue extends from the base of the stigma through the style and traverses the central septum separating the carpel ovaries. The style is a merger of two styles; thus there are two extended dorsal bundles, with a strand of transmitting tissue between them. At the base of the style the transmitting tissue becomes 4-lobed. At the top of the ovary, however, it changes shape to conform to the septum, and further down in the ovary it becomes rather diffuse. Pollen tubes in the ovary emerge from the septum through cracks in its thin cuticle and grow seemingly without direction until they contact a cluster of secretory cells (the obturator) near the entry to an ovule. The transmitting tissue in radish therefore guides pollen tubes close to, but not directly into, the ovules.

Citrus limon was mentioned earlier for its copious stigmatic secretion. The style has transmitting tissue that is intermediate between open and closed. It has several ribbonlike stylar canals filled with a material composed of lipids, polysaccharides, and proteins secreted by secretory cells that line the canals. The canals open downward into the ovary but do not extend upward to the stigma surface; instead, they end blindly in parenchyma tissue of the stigma. In tangelo and tangerine, also *Citrus* species, there is one wide central stylar canal surrounded by 10 or 11 narrow canals. The central canal ends blindly at the base of the style, whereas the others extend from stigma surface to the carpel ovaries (Kahn and DeMason, 1986). *Citrus* therefore provides an exam ple of considerable variation in transmitting tissue among species within one genus.

Red clover (*Trifolium pratense*, Fabaceae) was described by Heslop-Harrison and Heslop-Harrison (1982) as having yet another transmitting tissue variation. A single stylar canal extends from just below the stigma to the carpel ovary. It differs from that of *Citrus limon* in that the canal forms "lysigenously," which means that it is originally cellular but the cells degenerate later to form a stylar canal filled with the former cellular contents, which presumably help nourish pollen tubes. Similar lysigenous open styles have also been reported from other legumes. In a nicely illustrated example from the tepary bean, *Phaseolus acutifolius*, Lord and Kohorn (1986) showed a stylar canal open from ovary to base of the stigma, but not of lysigenous origin. Legumes therefore show considerable variation in their stylar canals.

True open styles have schizogenous canals formed by the nondestructive separation of originally compact living cells. These cells later line the open canal. Such canals extend through the stigma and style, opening directly to the surface at one end and to the gynoecial ovary at the other end. Compatible pollen tubes can therefore grow unimpeded from stigma surface to the micropyle of an ovule. Open styles are common in monocots, but the grass family is a curious and important exception because, although it has what must technically be called a stylar canal (really just a cleft), pollen tubes do not enter it but instead grow between elongate transmitting tissue cells that traverse the plumose stigma/style branches.

A surprising number of dicots distributed among many families also have an open style—for example, the familiar maples (dicot family Aceraceae) (Peck and Lersten, 1991). The gynoecium of black maple (*Acer saccharum*, ssp. *nigrum*), a subspecies of sugar maple, has two long papillose stigmas that converge to form a short open style lined with similar papillose cells. Secretions from these cells provide an unbroken pollen track from stigma to the obturator in the ovary, against which is pressed the micropyle of each ovule. Although this open style pathway is the obvious one, all pollen tubes observed by Peck and Lersten (1991) instead penetrated the stigma and grew internally between cells until emerging in the short style to grow superficially from there to an ovule. Nature is ever surprising, and this example should alert investigators to be cautious in making conclusions about pollen pathways without observing where pollen tubes actually grow.

The female flower of the kiwifruit vine (*Actinidia chinensis*, dicot family Actinidiaceae) has a conspicuous open stylar canal, as described by Hopping and Jerram (1979). The relatively wide canal is at first filled with very loosely organized cells that appear almost like long papillae. These papillae degenerate after pollination, leaving large gaps through which pollen tubes grow. The loose cells in the stylar canal and adjacent stylar parenchyma cells show a marked decline in starch grains while pollen tubes are growing. It could be argued that kiwifruit exhibits either a closed or open style, or one could make a case that both exist at different stages.

The way that transmitting tissue is arranged and connected has been considered by some to be crucial in determining whether a gynoecium exhibits true syncarpy even if the individual carpel ovaries are completely fused. Carr and Carr (1961) argued that true syncarpy must include a merger of all transmitting tissue strands, or at least a common junction where transmitting tissue from all carpels can meet, as in the maple example just described.

They argued that in true syncarpy a pollen tube germinating on any stigma must have an equal chance of entering any of the carpel ovaries. An example of true syncarpy is the carrot (*Daucus carota*, Apiaceae), which has five separate styles and two carpel ovaries, but the transmitting tissue from each style merges into a common pad of tissue at the style base before separating again into one branch for each ovary. This common junction allows a pollen tube from any style to randomly enter either ovary.

Carr and Carr (1961) called this transmitting tissue junction a "compitum" (Latin for "a crossing place"). A species such as apple lacks a compitum, and each carpel is therefore independent with respect to the path followed by individual pollen tubes. Carr and Carr were not specific about the advantages of a compitum beyond mentioning that the fruit in such species has a good chance of developing symmetrically because at least some ovules in all of its carpels will be fertilized and produce seeds. Endress (1982) suggested that the real advantage of a compitum lies in the increased competition between pollen tubes as they converge in the compitum. This competition should result in the selection of only the most vigorous pollen tubes, because all tubes must cross the same finish line, so to speak; therefore, only the fastest growing ones win an ovule as a prize.

Carr and Carr (1961) also commented on transmitting tissue in the carpel ovary, about which information is rather scanty. They reported that their observations on a considerable but unstated number of dicots and monocots always showed a cuticle on the internal surface of the ovary; transmitting tissue formed a pollen track along the inner wall of the ovary, and the cuticle over this track was thinner than elsewhere. Secretions either seep through this thin cuticle or they accumulate below it and lift it, causing a rupture. Pollen tubes grow along such pollen tracks to the ovules. A similar example of this type of track was described in the ovary of tepary bean, in which the two parallel placental ridges that act as pollen tracks lack a cuticle, whereas it is present elsewhere on the inner ovary epidermis (Lord and Kohorn, 1986).

Transmitting tissue is also known by other terms, either collectively for its entire length in the carpel, or by separate terms for the stigma, style, and ovary. Tilton and Horner (1980) reviewed the subject, for which there is published information extending surprisingly back to the late 1700s. Based on their review, and on their detailed study of the entire gynoecium of soap plant, *Ornithogalum caudatum* (Liliaceae), transmitting tissue in all parts of the gynoecium appears to have the same morphological origin. Therefore they support the use of the common term "transmitting tissue," as qualified by stigmatic, stylar, and ovarian where appropriate. They regard the obturator, a usually conspicuous outgrowth that occurs next to the ovules in some families, as a special type of ovarian transmitting tissue that is closely associated with the micropyle of the ovule. The physiological aspects of the style and stigma are taken. Style and stigma development before pollination is not as dramatic as many of the examples of rapid stamen elongation already mentioned, but their formation and maturation have also been shown to be related to the action of hormones. The small number of studies, however, indicate differences in response to various plant hormones among species. Koning (1983) showed that the interaction of three different hormones (gibberellin, auxin, ethylene) controls style and stigma growth in the garden ornamental, gaillardia *Gaillardia grandiflora*, (Asteraceae).

6

Ovule and Embryo Sac

An ovule ("little egg" in Latin) is a megasporangium, the equivalent of a microsporangium (pollen sac). It arises as a tubercle from the inner wall of the ovary in either a parietal or axile location. If it develops normally, if a pollen tube enters it and releases compatible sperm, and if normal fertilization and post-fertilization events ensue, the ovule will eventually become a seed.

There can be from one to several hundred ovules in an individual carpel ovary, and the entire gynoecium (all carpels together) of a single flower may bear thousands of ovules. Except for the obviously necessary egg cell, much of the rest of the ovule seems strange at first, and therefore it should be described in some detail before moving on to how it relates to the pollen tube and fertilization.

This chapter deals mostly with ovule development, mature structure, and functional aspects, including the embryo sac (female gametophyte) that is retained within it.

These topics have intrinsic interest, but they also provide the necessary background, which deal successively with pollination through fertilization, endosperm, and embryo, all of which intimately involve the ovule.

Certain aspects of seeds, including seed failure, are also included in this chapter because they seemed to belong best here. For a detailed morphological and developmental account of ovules to seeds, consult Bouman (1984).

OVULE FORM AND DEVELOPMENT

Previous chapter points out that the carpel in the earliest angiosperms probably resembled a simple leaf folded longitudinally or rolled into a cylinder. These seemingly simple developmental maneuvers provided a protective enclosure for the ovules, which occurred in two longitudinal rows, one along each side of the suture joining the margins. In most modern families, however, the gynoecium consists of multiple carpels that adhere or merge with each other in any of several possible and often difficult to interpret ways during floral development.

How carpels unite (or remain separate) determines whether ovules will be attached to the inner peripheral wall of the carpel ovary, to a central axis or free-standing post, or arise from the base of the carpel ovary. Carpel mergers can also affect how many ovules will occur. A typical ovule consists of three parts: nucellus, funiculus, and integuments. The central body of the ovule is called the nucellus (Latin for "a small nut," conveying the meaning that this is the core of the ovule). Within the nucellus, one cell—the megaspore mother cell—enlarges and undergoes meiosis to produce four megaspores, of which only one (the functional megaspore) continues to develop into the embryo sac.

The base of the nucellus commonly tapers into a stalk, called the funiculus (Latin for "a small rope"), which attaches the ovule to the placenta of the carpel wall. In most ovules the funiculus is bent back on itself 180 degrees or at least 90 degrees, so that the nucellus is recurved to face the placenta, often pressing against it.

The integuments (Latin for "covers") are two superimposed cuplike sheaths (inner and outer) that arise from the base of the nucellus in a vaguely defined area known as the chalaza (Greek for "hailstone"—the sense of this is obscure). The integuments encircle the nucellus and at least one of them overtops it, leaving only a small apical slit or pore. The second integument (either the outer or inner one) may develop to the same degree, may extend only partway up the side of the nucellus, or in some species may abort early or never form at all. The apical opening of the outer integument is called the exostome ("outer mouth"), and that in the inner integument is the endostome ("inner mouth"); together they comprise the micropyle ("a small gate"). Exostome and endostome may coincide to form a straight micropylar entryway onto the nucellar surface, or they may be offset from each other above the nucellus, which requires the entering pollen tube to bend twice in order to reach the nucellus. This offset arrangement, which for example is common in legumes, is called a zigzag micropyle.

The pollen tube must grow through the nucellus to reach the embryo sac and release its sperm cells, but in some plants it does not grow through the micropyle as just described, but instead enters in what seems the hard way, growing in from the flank or base of the nucellus. In a walnut ovule, for example, Nast (1935) reported that a few tubes enter through the micropyle but most grow through the base of the nucellus. Since the micropyle in many species is not an easily entered open gap, perhaps entry elsewhere is not radically different. At the other extreme are plants in which the embryo sac expands and destroys the overlying nucellar layers and pushes its way partly or completely through to the surface of the micropyle. This behavior occurs, for example, in sunflower and African violet, which exposes the embryo sac directly to a pollen tube at the tip of the ovule.

Philipson (1977) surveyed dicots and reported that bitegmic (two integuments) ovules are found in three times as many families as are unitegmic (one integument) ovules. Why two integuments should typically develop instead of one is unknown; it is also puzzling that some ovules that start out bitegmic end up unitegmic because the inner integument is either crushed or stops growing early and disappears. Two unitegmic families of dicots with many cultivated species are the carrot family (Apiaceae) and the mint family (Lamiaceae).

To complicate matters, a third integumentlike ovular outgrowth, the aril (Latin, meaning raisin or grape seed—the sense of this is obscure), occurs in several unrelated families. Arils may be colored and/or thick and fleshy, and may cover the mature seed. Arils contain edible substances such as oils, an adaptation that attracts animals to carry them away for dispersal. The dried aril of the nutmeg seed from which the spice called mace is processed is an example of an aril that has economic importance. Most ovules are bent back on themselves 180 degrees or nearly so, usually because of curvature of the funiculus.

By this developmental maneuver, such an anatropous (meaning approximately "to turn oppositely") ovule bends its micropyle conveniently close to the placenta to provide a convenient entry for pollen tubes growing along the inner carpel wall. Anatropous ovules are known in over 200 families and, if one includes ovules that bend back only to somewhere between halfway and completely—e.g., in spinach— probably more than 75% of angiosperms have ovules that turn back to a lesser or greater degree toward the placenta. Curvature of 180 degrees also presses the nucellus against the funiculus, and the two may adhere or actually merge during development. This has occurred, e.g.., in bergamot. In a fully developed anatropous ovule, the funiculus may therefore be indistinguishable as a separate structure.

In contrast to an anatropous ovule, or even to one bent halfway back, there seems to be no easy pathway for pollen tubes to enter an atropous (meaning "without turning") ovule, one that protrudes straight out from the placenta and positions the micropyle toward the interior of the carpel. Atropous (also called orthotropous) ovules occur in only about 20 families, including familiar nut tree species such as walnut and pecan (both Juglandaceae). Atropous ovules also occur in some mostly herbaceous families, such as Polygonaceae, of which rhubarb and buckwheat are members.

Most of the 20 atropous families have ovaries with a single large ovule. Such ovaries may secrete a chemical attractant or have some physical means (i.e., the ovary might fit tightly around the ovule) to channel pollen tubes to the micropyle, or pollen tubes might enter the base or flank of the ovule, as is common in walnut, for example (Nast, 1935). Legumes have a bitegmic ovule in which the nucellus as well as the funiculus is curved, and the embryo sac within the nucellus is also curved, as is the mature embryo in the seed. This variant of the anatropous ovule is termed campylotropous (meaning "curved turning"). Campylotropous ovules also occur in some other families, e.g. Chenopodiaceae. Ovule curvature is quite variable among grasses, where species with atropous, anatropous, and intermediately curved ovules all occur. A distinguishable funiculus either does not occur in grass ovules, or it is inconspicuous. In maize, for example, the bulky ovule seems almost atropous, although the micropyle is oriented away from the stylar cleft, and its two integuments are quite asymmetrical.

The development and mature form of a representative bitegmic, anatropous ovule can be seen in an excellent series of scanning electron micrographs of spiral flag, *Costus cuspidatus*, an ornamental monocot of the ginger family (Zingiberaceae). The integuments gradually overtop the nucellus while the elongating funiculus recurves, which eventually bends the ovule 180 degrees back on itself. The micropyle ends up against the placenta, where a pollen tube can easily enter. Perhaps because of this hairpin curve, that part of the outer integument squeezed between the nucellus and the funiculus either fails to develop, or else fuses with the funiculus. As the developing embryo sac expands, the surrounding nucellar tissue also proliferates; thus the nucellus maintains itself, and perhaps even becomes enlarged, in the mature ovule.

In some angiosperms—for example, the group of dicot families informally called Centrospermae— the nucellus proliferates into a tissue called perisperm (meaning "around the seed") as the embryo develops. Nucellar perisperm either provides nutrients that supplement the endosperm, or in some species even replaces the endosperm. Spinach (Chenopodiaceae) provides an example.

Perisperm also occurs among monocots, for example in *Agave* (Agavaceae). Why the typical endosperm (which is a triploid or polyploid tissue) should be replaced by diploid nucellar perisperm tissue for embryo nutrition in some plants is unknown. The ovule of cultivated flax (*Linum usitatissimum*, dicot family Linaceae) is of the common bitegmic, anatropous type. The micropyle is not an open passage but is instead a long fissure (endostome) through the thickened inner integument, which appears to be a tight squeeze for a pollen tube. The exostome would be a wide opening except that both integument tips are pushed against an obturator, which secretes a substance that presumably attracts the growing tip of any pollen tube in the vicinity. If this figure is an accurate depiction of the receptive ovule, the pollen tube does not have easy entry. Expansion of the developing flax embryo sac, along with growth of the inner integument, has obliterated the nucellus. Nucellar destruction is common in ovules of many families.

The conspicuous layer of radially elongate cells surrounding the embryo sac in flax is the inner epidermis of the inner integument. Such an integumentary tapetum or endothelium usually occurs in ovules where the nucellus has been crushed by embryo sac expansion, and as a result the naked embryo sac is surrounded directly by the inner integument. Published descriptions of the integumentary tapetum were compiled and reviewed by Kapil and Tiwari (1978), who found reports from 65 dicot families but, surprisingly, no reports from monocot families. The enlarged inner integument epidermal cells often have dense cytoplasm, and callose deposits have been reported in some species. The function of the integumentary tapetum is unknown, but they speculated that it transfers nutrients to the embryo sac and that it might contribute to the protection of the immature embryo as the seed develops.

In contrast to flax, the bitegmic anatropous ovule of the grape (*Vitis vinifera,* dicot family Vitaceae) retains its relatively massive nucellus and there is no integumentary tapetum. The micropyle is straight and open, but it is formed only by the endostome since the outer integument does not grow over the top of the nucellus.

An ovule is typically supplied by a single vascular bundle that branches from a ventral bundle, traverses the funiculus, and ends in the chalaza of the nucellus. The chalazal cells that lie just beyond the end of the ovular vascular bundle may enlarge, develop a conspicuous cell wall, or accumulate dense contents. If any of these modifications occur this area is referred to as a hypostase (Greek for "foundation," evidently a reference to the base of the nucellus between embryo sac and funiculus). Tilton (1980) reviewed published information on variations in the hypostase and concluded that its primary function is probably to aid movement of nutrients into the developing embryo sac and embryo; as a secondary function, the hypostase may become a storage tissue. A single ovular vascular bundle that traverses the funiculus and ends at the base of the nucellus is most common among angiosperms, but in some families this bundle splits into several smaller ones that extend variously into the outer integument, but only rarely into the inner integument. Such vascular penetration into one or both integuments has been reported in about 80 families, including Cucurbitaceae, Rosaceae, and Fabaceae. In some legumes, such as peas and soybean, these terminal vascular bundles have been shown to transport nutrients into the integuments, which probably hastens seed maturation (Murray, 1987).

OVULE FAILURE AND OVULE ABORTION

Can all of the ovules in a flower develop into viable seeds? Numerous observations and investigations concerning this question have shown that it is typical for at least some ovules to abort at any stage before seed maturity. This can occur for three possible reasons:

- Too many ovules are produced.
- Accidents occur during ovule development and fertilization.
- Some ovules in certain species are programmed to abort.

Overproduction of ovules exceeds the capacity of the gynoecium, and by extension the whole plant, to provide adequate nutrition for all of them. Some studies have shown that where nutritional resources are finite, which is a common situation for many reasons, ovules that were fertilized first, up to a certain number, draw off the available nutrients and leave laterfertilized ovules to starve and abort. The hypostase and integumentary tapetum described earlier might be involved in this. Haig and Westoby (1988) felt that both structures probably channel nutrients into the embryo sac, but they can also be converted into barriers to nutrients. At some stage in ovule development these cells may develop callose or tannin deposits, or otherwise modify their wall structure, changes that Haig and Westoby interpreted as evidence that these cells have the ability to " . . . also seal off the seed from maternal resources." They speculated that the maternal parent thereby selectively allocates its resources to certain ovules and causes others to abort.

Second, any ovule may abort because any one of the many steps in normal reproduction on both male and female sides can go wrong even under the best conditions, just as it can in all animals and certainly among human beings. A variable number of ovules will therefore abort because of some accidental error in development or fertilization.

Abortion because of these two reasons has been documented by many careful published studies that include data on the number and percentage of ovules that become viable seeds. A detailed consideration of these studies is mostly outside the scope of this book, in part because such studies are typically non-embryological in nature. But a few examples can provide insight into what commonly happens in the gynoecium of cultivated plants. The chickpea, *Cicer arietinum*, only produces 15–50% of its potential number of seeds (Bassiri et al., 1987), which means that half or more of the ovules abort at some stage. And in the avocado, which has only one ovule per ovary, only a small percentage of flowers typically produce seed even when sufficient pollen is provided by hand (Sedgley, 1979).

A careful study by Pechan (1988) showed what happens in rapeseed (*Brassica napus*) even under the best conditions. Using gross observational methods coupled with detailed views from sectioned ovaries, Pechan showed that several kinds of "accidents" caused some of the ovules in an ovary to abort, even when surplus pollen was placed on the stigma to provide more than enough pollen tubes. He reported that some ovules were simply bypassed, some were incompletely penetrated, and in others fertilization occurred but subsequent development went awry and the seed aborted very early. Still other ovules were fertilized, appeared to develop normally, but seemingly could not compete successfully for nutrients and thus remained small and eventually aborted.

A study by Al-Jaru and Stösser (1983) followed pollen tubes and the fate of ovules in red currant (*Ribes rubrum*) and black currant (*R. nigrum*) shrubs (dicot family Saxifragaceae). They found that even with optimal growing conditions and plenty of compatible pollen provided, many pollen tubes reaching the ovary either appeared abnormal or grew haphazardly and did not penetrate an ovule.

The third and most mysterious way that ovules are eliminated is by programmed abortion, which can occur either before fertilization or shortly thereafter. Maples (*Acer*, dicot family Aceraceae), for example, have four ovules per flower but only one develops into a seed. The female oak flower (*Quercus*, family Fagaceae) has five ovules; most of them will develop normally and be fertilized, but four ovules will invariably abort early in seed development and only one will reach maturity (Mogensen, 1975). Mogensen speculated that the first fertilized ovule is able in some unknown way to suppress the further development of any other ovules. A similar process may occur in maples.

Programmed ovule abortion occurs among cultivated species of *Prunus* (Rosaceae: cherry, plum, almond, and other edible fruit species). The simple gynoecium of these species has only two ovules, of which one always aborts before fertilization. Pimienta and Polito (1982) showed how this occurs in the almond, *Prunus dulcis*. The aborting ovule either lacks an embryo sac or else it forms late and develops abnormally. Callose appears two days after pollination in the chalaza of the aborting ovule, spreads into the inner integument, and eventually encloses most of the nucellus. Treatment with a fluorescent dye showed blockage of the ovular vascular bundle at the chalaza, indicating that callose prevented entry of water and nutrients, whereas the dye flowed unimpeded throughout the callose-free normal ovule. Self-regulated abortion of some ovules perhaps involves hypostase and endothelium blockage, as speculated by Haig and Westoby (1988).

MEGASPOROGENESIS

The formation of megaspores in the ovule has parallels to microspore production in the pollen sac because meiosis occurs only in a special cell (megaspore mother cell), which subdivides itself into four megaspores, each with a nucleus containing half the number of chromosomes that were present in the parent cell. The stages of meiosis in megasporogenesis reportedly progress similarly to meiosis during microsporogenesis (Bennett, 1977). Time required for meiosis is also similar, although only about six species have been studied in detail, as compared to 39 species for microsporogenesis. There are, however, two conspicuous and important differences in the ovule:

- There is only one megaspore mother cell per ovule (exceptions are known, but these are mostly abnormal events) compared to several to numerous microspore mother cells per pollen sac.
- Only one haploid megaspore survives after meiosis, compared to four microspores from meiosis of the microspore mother cell. A complication, however, is that the functional megaspore may have one, two, or four nuclei.

The typical situation in an ovule is that a single cell, which could be considered a 1-celled archesporium, enlarges and becomes the megaspore mother cell, and then undergoes meiosis. Normal meiosis produces four cells, or at least four nuclei, so a mechanism must exist to reduce the four meiotic products down to one, in order to end up with a single megaspore. Because it and the female gametophye it will become after internal germination are destined to remain in the ovule to nurture the embryo, one is as many as a normal ovule can support.

More than a century of observations on thousands of species have demonstrated three possible ways that plants handle the problem of going from four to one. These are the three types of megasporogenesis, which are shown 92 *Flowering Plant Embryology*. The monosporic type is most common, in which meiotic cytokinesis is successive, producing four megaspores.

Three of these megaspores degenerate, and one, which is almost always the deepest (chalazalmost) one, lives on to become the functional megaspore. The bisporic type differs because one of the two cells formed by cytokinesis of the first meiotic division degenerates; after the second meiotic division no cytokinesis occurs, and the resulting functional megaspore is binucleate; the term bisporic refers to the two nuclei that would have been in separate spores if a wall had formed. Meiosis in the tetrasporic type lacks cytokinesis and all four haploid nuclei remain within the 4-nucleate megaspore; tetrasporic refers to the nuclei that would have been in four separate spores if cytokinesis occurred after each meiotic division. The bisporic and tetrasporic pathways to megaspore formation are not common, but as examples one can point to the onion (bisporic) and the cultivated lilies and their relatives (tetrasporic).

In the ovule, as in the anther, callose has been detected during megasporogenesis, a phenome non described most comprehensively by Rodkiewicz (1970). The distribution of callose deposited during megasporogenesis resembles that deposited during microsporogenesis, but there are some differences. The general pattern in the ovule is that callose in the monosporic type surrounds all four megaspores at first, and sometimes even the megaspore mother cell before meiosis, but later it dissolves only from around the functional megaspore. In some species, however, callose never forms around the functional megaspore. Thus it appears that further development requires early dissolution of callose, whereas in the pollen sac it is normal for callose to be retained for some time after meiosis. Observation of callose during megasporogenesis is unfortunately documented by far fewer examples than from pollen sacs.

Three examples, two from dicots and one from a monocot, serve to illustrate megasporogenesis of the common monosporic type. A detailed series of drawings provided by Bushnell (1936) for the cultivated mint bergamot (*Monarda fistulosa*, Lamiaceae) shows the process with textbook clarity. The single megaspore mother cell enlarges and undergoes the first meiotic division, which produces a large chalazal cell and a smaller micropylar cell. These two cells then each undergo the second meiotic division, resulting in three small micropylar megaspores that degenerate, and one large chalazal functional megaspore, which is shown in position in the ovule, where it dominates the nucellus. Monosporic development with a variation was shown by Kennell and Horner (1985) in the soybean ovule. A young whole ovule prepared by a special clearing procedure reveals the single enlarged and centrally located megaspore mother cell in the nucellus, which is partially covered by the developing integuments. After meiosis only the chalazal megaspore survives and develops further. Degeneration of megaspores can begin even before all four megaspores are completely formed. The same figure, seen after special staining and under fluorescence microscopy, shows callose associated with the cross-walls of the degenerating megaspores. At a later stage (not shown) only the chalazal megaspore survives.

A similar sequence from a monocot can be seen in maize, as described by Russell (1979). A longisection of a young ovule shows the bulky nucellus with one large and conspicuously elongate megaspore mother cell in the center. It is already surrounded by callose before meiosis, as shown by fluorescence microscopy. Meiosis results in a linear tetrad of megaspores as in soybean, but in maize the two micropylar megaspores often begin to degenerate so quickly that they appear as one cell, and thus a triad of megaspores is seen instead of a tetrad. A similar triad specially stained and photographed using fluorescence microscopy shows that callose persists around those megaspores that will degenerate, but it is now absent from around most of the functional megaspore. Callose remains around the degenerated megaspores even after the functional megaspore has begun to enlarge and develop into the embryo sac.

Callose secretion during megasporogenesis shows similarities to callose secretion during microsporogenesis. The failure of callose to dissolve from around a tetrad of microspores at just the right time is a common cause of microspore and pollen abortion in certain kinds of cytoplasmic male sterility. The failure of callose to dissolve from around three of the four megaspores could also be the direct cause of abortion. There is a difference, however, because prolonged retention of callose is normal behavior in the ovule but it is abnormal in the pollen sac.

The hypothesis that persistent callose causes megaspores to abort is strengthened by two additional observations. The first is that callose in many members of the evening primrose family (Onagraceae) dissolves from around the megaspore closest to the micropyle, which is the opposite of callose dissolution in other families; thus the micropylar megaspore is functional instead of the chalazal megaspore. The second observation, which is unfortunately based on only a few examples, is that no callose has been seen to form in the tetrasporic type of megasporogenesis, the type of development in which the four nuclei from meiosis are crowded into the sole megaspore, which is the functional one.

Callose has been reported to be deposited incompletely, or even omitted entirely, from around the functional megaspore in some species. This supports the hypothesis that callose around the other megaspores contributes to their degeneration, but it weakens the hypothesis that callose must be present to isolate the haploid products of meiosis from surrounding sporophyte tissue, as proposed for microsporogenesis. It also supports an alternative hypothesis that a callose sheath prevents entry of viruses or other types of invasive pathogens (Heslop-Harrison et al., 1999). More observations are needed, as well as some experimental work, before a generalization can be formulated concerning how, or if, the pattern of callose deposition and dissolution in the ovule has a different function from that in the pollen sac.

It appears that callose non-formation or dissolution at the right time coupled with other unknown factors combines to select the chalazal megaspore to continue on in the great majority of monosporic and bisporic species that have been studied. Why should the deepest- lying megaspore be in the favored position? One could argue that the distances separating chalazal from micropylar are so tiny that it should not matter, in a positional sense, which one is selected, and therefore one might expect much more variability. The evening primrose family exception (and a few others as well) seems to prove that a micropylar functional megaspore can fare just as well. No answers have appeared as yet.

One final topic concerning megasporogenesis that deserves mention is the location of the megaspore mother cell in the nucellus. The illustrations of bergamot, soybean, and maize cited earlier in this chapter showed this cell just beneath the nucellar epidermis, a location called tenuinucellate ("thin nucellus"), as in Figure 6.6. A megaspore mother cell and its resulting embryo sac deeply embedded in the nucellus is called crassinucellate ("thick nucellus"), as in the grape. The difference between a superficial embryo sac and a deeply embedded one is probably related in some way to seed structure, and its significance must therefore be sought in the individual species and not as a grand generalization. But it seems likely that a pollen tube confronting a crassinucellate embryo sac must find that it presents a challenge, unless it has special enzymes to use, whereas tenuinucellate embryo sacs are close to the nucellar surface and should be much easier to penetrate.

EMBRYO SAC (MEGAGAMETOPHYTE) DEVELOPMENT

Following megasporogenesis the common monosporic type retains only one surviving megaspore with one nucleus. Plants with bisporic development have one binucleate megaspore; in tetrasporic types there is no degeneration and the sole megaspore has four nuclei. Each of these types of functional megaspore remain embedded in the nucellus and therefore must germinate within itself in order to develop into a female gametophyte (embryo sac). This parallels pollen development because a microspore also germinates internally to form the male gametophyte (pollen grain) while still within the pollen sac. Both microspore and megaspore behave similarly in that sense, but the megaspore produces a larger and more complex entity, which is appropriate since it must later contribute to the formation and nurturing of both embryo and endosperm.

A minimal embryo sac could be imagined, one that is reduced to only two cells (egg cell and uninucleate primary endosperm cell). Although no embryo sac that simple has ever been described, some species are known with a reduced embryo sac consisting of only four cells (3-celled egg apparatus and primary endosperm cell), which barely surpasses a tricelled pollen grain.

After more than a century of research on thousands of species, 10 different developmental pathways from megaspore to embryo sac are known: two monosporic, one bisporic, and seven tetrasporic variants. Some genetic variability is involved in the number of meiotic nuclei that participate in some of the embryo sacs, but since the end result is always an embryo and endosperm, it is difficult to attach great significance to these different pathways to the same end. Because the megaspore germinates within itself, the embryo sac must grow within the megaspore wall, which must expand to accommodate the internal growth. The literature reveals mostly only casual and qualitative statements about this aspect of its growth.

Heslop-Harrison et al. (1999) did mention various structural changes in the wall during development, which they interpreted as providing physical protection against entry by pathogens. They proposed that the callose sheath around the megaspore also provides protection from pathogens. George et al. (1988) are almost the only investigators to provide some statistically valid observations on rates of embryo sac enlargement. They reported that growth rates in two legumes, soybean and mung bean (*Phaseolus aureus*), differ during certain stages of development, and that these growth rates are in turn affected by whether plants have been growing in field or greenhouse. For example, embryo sacs in field-grown soybeans increased in length most rapidly between 2- and 4-nucleate stages, whereas in the greenhouse the greatest increase occurred between the functional megaspore and 2-nucleate stage. The embryo sac wall (i.e., the former functional megaspore wall) must expand to accommodate this internal growth.

The environmental influences noted by George et al. (1988) on the rate of embryo sac growth can also affect the pattern of nuclear division and arrangement of cells within the embryo sac. The embryo sac does not always have to develop only one way in a species; variations within one species were cited by

Hjelmquist (1964) for several different species. In some elm species (*Ulmus*, dicot family Ulmaceae), any one of three different patterns of development may occur (Hjelmquist and Grazi, 1965). These variations were attributed to the effect of temperature fluctuation at critical times during embryo sac development. But whatever variation occurred, these workers reported that normal seeds always resulted.

One can speculate that variant patterns of embryo sac development probably occur from time to time in other species as well, without necessarily affecting normal seed development. Very little statistical sampling within an individual species has been reported in the embryological literature but, extrapolating from the large number of angiosperm species investigated over more than a century, it is evident that the vast majority, including most cultivated species, conform to one pattern, the so-called "normal" or "*Polygonum*" type (named for the genus in which it was first described).

Three of the other variants shown in Figure 6.15 also end up with the same final pattern of cells and nuclei. The cultivated lily is almost certainly the most extensively sampled species that exhibits uniformity of embryo sac development because it is the commonest species used to demonstrate megagametophyte development. The lily embryo sac consistently conforms to the tetrasporic *Fritillaria* type.

The *Polygonum* or normal embryo sac is initiated by one megaspore with one nucleus, as described in the earlier examples from wild bergamot, soybean, and maize. This monosporic beginning can be continued using maize as a representative species: three mitotic nuclear divisions produce a coenocytic embryo sac containing eight nuclei within the original but greatly expanded functional megaspore wall. The embryo sac is therefore unlike the pollen grain because it passes through a coenocytic phase. Some bisporic and tetrasporic species follow the same pattern, but they only require two and one nuclear divisions, respectively, to reach the 8- nucleate state; their nuclei, unlike the monosporic species, are not all genetically identical. The embryo sac continues to enlarge as these nuclei divide, and in some plants it also elongates considerably, as in silver maple and many cultivated members of Rosaceae.

In the almond (*Prunus dulcis*, Rosaceae) Pimienta and Polito (1982) showed that a thick, callose-like shell appeared around the elongating embryo sac and remained until after fertilization before it disappeared. They mentioned that callose has also been reported to occur later, surrounding the embryo sac in *Petunia*. They suggested that a coating of callose might be necessary to isolate the female gametophyte from the surrounding sporophyte tissue, a speculation that has also been made for the several other places in both microgametophyte and megagametophyte development where callose appears, but they prudently advised that more examples need to be verified before an explanation is ventured with respect to the embryo sac.

The eight nuclei in the embryo sac arrange themselves in two clusters of four nuclei each at opposite ends. One nucleus from each end migrates toward the middle and these so-called polar nuclei, named for where they came from rather than where they end up, often remain as a pair in an intermediate position in the embryo sac. In some plants the polar nuclei merge before fertilization, as in silver maple, a technically difficult phenomenon to study (and therefore little known) that is probably very similar to the merger of egg and sperm nucleus during fertilization. Whether separate or fused, polar nuclei do not have to be centrally located but may instead take a position close to, or even abutting, the egg apparatus.

How these eight nuclei move around is unknown, although there are several reports from the literature that remnants of spindle fibers from earlier nuclear divisions may be involved. There have been suggestions that the two synergid nuclei come from a common mother nucleus and that the egg nucleus and the micropylar polar nucleus are sister nuclei (Cass et al., 1985, 1986).

Six of the eight nuclei, each with some surrounding cytoplasm, become isolated from each other by new cell membranes that form individual cells, but with only partially complete cell walls. An embryo sac of the normal type therefore consists of eight nuclei distributed among seven cells or seven nuclei if the two polar nuclei merge before fertilization. The large binucleate central cell is really the enlarged original megaspore, which has within itself a trio of cells at each end. A complication is that the antipodal cells may quickly degenerate or they may proliferate into many cells, which means that the normal type may vary from the configuration just described. More about this phenomenon appears later in this chapter.

Here it is possible again to see a parallel with the pollen grain. The generative cell in a pollen grain lies entirely within the nutrient-rich vegetative cell. The common type of female gametophyte has six cells that lie within the central cell of the original megaspore. After fertilization the central cell accumulates nutrients and becomes endosperm, and in that sense can be compared to the nutrient-containing vegetative cell of the pollen grain.

The chalazal trio of cells in the embryo sac are called antipodals (Latin, "against the foot"), referring to their location at the opposite end from the essential three-celled egg apparatus at the micropylar end. The egg apparatus consists of the larger egg cell flanked by two smaller cells called synergids (Greek for "helpers" or "cooperators"), all of them reportedly with an incomplete cell wall. The synergid cell wall in contact with the original megaspore wall, which is now the embryo sac wall, is thickened and spongy in appearance, with numerous cytoplasmic channels. This specialized portion of the synergid wall is called the filiform apparatus.

The large binucleate cell that contains the other six cells is the original megaspore protoplast, which is called the central cell. The central cell nuclei are usually called polar nuclei, a term that does not indicate nuclear position, but rather that each nucleus came from an opposite pole.

To continue with the interrupted story of *Polygonum* type of development in maize. Note that the original three antipodal cells have now proliferated to several, a seemingly pointless but perhaps useful increase in number, which is common among grasses and certain other families (e.g. mustard family among dicots); increases to a surprising 100 or more antipodal cells occur in some species.

Observing when and how cell walls form within the embryo sac to partition it into antipodal cells at the chalazal end and three egg apparatus cells at the micropylar end has been very difficult because the walls are thin or incomplete, and they form quickly. A combined light- and electron-microscope study of wall formation within the embryo sac of barley by Cass et al. (1985, 1986) revealed wall formation among the egg apparatus cells. They found that only two cell plates, each similar to those of dividing vegetative cells, delimit the three cells, which is possible because one cell plate branches and curves. A cell membrane always surrounds each cell, but no cellulose wall may form or it may form only in patches, a temporary omission that seems necessary to allow fertilization to occur.

The odd *Polygonum* type of angiosperm embryo sac is not the end point of embryo sac evolution, as shown by the occurrence of several variations (Fig. 6.15). One surprising variation is the *Plumbagella* type, a reduced embryo sac of some members of the dicot family Plumbaginaceae, including leadwort (*Plumbago*), a warm-climate ornamental shrub. In this type both synergids are lacking, but the specialized filiform apparatus has instead become part of the egg cell wall, which thus combines in one cell the functions of both egg and synergids.

CELLS IN THE NORMAL (*POLYGONUM*) TYPE OF EMBRYO SAC

Among the three cells of the egg apparatus the egg cell has an obvious purpose. It is larger than the synergids, and it has either no cell wall or only a patchy wall until after fertilization. More about this cell in Chapter 8, where fertilization is discussed.

The function of the accompanying pair of synergids, which also have an incomplete cell wall, was long considered to be questionable. Many earlier investigators detected no activity related to fertilization, and one or both synergids had been variously reported to degenerate before or during pollen tube entry into the embryo sac. It had also been reported that a synergid was sometimes penetrated by the pollen tube, sometimes not. In view of such seemingly meaningless synergid behavior, Maheshwari (1950) concluded "This seems to indicate that they are not essential for fertilization."

More recent investigations using electron microscopy have verified that the pollen tube always enters one synergid through the filiform apparatus, although it should be mentioned that there are not many such studies. There is some evidence that the filiform apparatus secretes a substance that attracts the pollen tube to the micropyle. It is fair to ask, if only one synergid is functional, why are there two of them? A pair might exist because it is a matter of some necessary symmetry of development, two cells might be needed to provide enough attractant for a pollen tube, or it might be that two synergid cells have been retained from an earlier stage of angiosperm evolution when each had a function. The problem of the many reports of pollen tubes that enter elsewhere than through the micropyle also needs to be resolved.

At the opposite end of the embryo sac are the antipodal cells. In many species they remain as a trio of cells, but in others they degenerate even before fertilization—as, for example, in most legumes and in buckwheat. In sharp contrast, antipodal cells can also proliferate in some plants, as mentioned earlier for grasses and mustards. Antipodal cells among members of the carrot family (Apiaceae) show a remarkably wide spectrum of behavior, ranging from seemingly functionless pre-fertilization degeneration to retention of the three cells to proliferation of a dozen or more cells (Gupta, 1964). Antipodal cells proliferate even before fertilization in some species, which means that the *Polygonum* type of embryo sac sometimes consists of more than the seven cells commonly described for it, and antipodal proliferation may affect embryo sac cell number in other types as well.

In addition to a considerable range of cell numbers, the nucleus of individual antipodal cells might become highly polyploid and/or the cell wall might develop transfer cell wall ingrowths. Such features are also known from cells in other parts of plants, where they are associated with heightened metabolic activity. It has been suggested that the antipodal cells in wheat help transport nutrients rapidly to the endosperm during its early, coenocytic phase of development (Bennett et al. 1975). It is evident that antipodals vary so much among species that no generalization about them can be made with any confidence.

The central cell is binucleate but why it needs two nuclei is still not explained adequately. In fact, that the monosporic *Oenothera* type of embryo sac has only one central cell nucleus and that some of the other types of embryo sacs have more than two. Speculation is that after fertilization the typically triploid nucleus enables the central cell to preferentially attract nutrients by some as yet unexplained means, which enables this cell to develop further as endosperm.

In many species starch grains accumulate in both central cell and egg cell before fertilization (Wardlaw, 1965). Brink and Cooper (1940) listed several examples of starch accumulation, especially in species of grasses and legumes, among the 37 mostly cultivated species that they examined. Soybean is an example. Another feature of the central cell is the development of wall ingrowths, which has the effect of extending the cell membrane, thus increasing its surface area and allowing more of certain molecules to be transported across it per unit of time. These transfer cell walls have been reported for several species in recent years as embryo sacs have been investigated by electron microscopy, and they are now known, for example, in such cultivated plants as cotton, flax, soybean, and sunflower, among others.

7

Pollination and Pollen-Stigma Interaction

Pollen develops in a fluid medium, and the pollen sac interior probably dries out only just when the anther dehisces. Following dehiscence the pollen grains, except for those in some aquatic species and species with permanently closed flowers, are exposed to drier outside air and carried away by insects, other animals, or air currents. These grains can survive for only a short time, from less than a day to a few days at most. A grain that quickly reaches a receptive stigma of the same species begins rehydrating and exuding substances (especially proteins) that react with molecules on the surface of receptor cells on the stigma or with extracellular substances (also mostly proteins) in stigmatic secretions. Favorable mutual recognition stimulates germination, and a pollen tube emerges and grows normally. A pollen grain on an incompatible or nonreceptive stigma, in contrast, is either not recognized or is actively inhibited by the stigma, and will not germinate. Alternatively, it will send out a tube that stops growing on the stigma, stops growing after growing only a short way into the style, or—more rarely—stops growing after it traverses most or all of the style.

This chapter describes how pollen reacts to the extreme environmental shifts from anther to air to stigma, the life span of pollen, how stigmas "capture" pollen, and the possibilities for successful or unsuccessful interaction between pollen and stigma or style. Left mostly out of consideration is pollination biology, a large and complex discipline that deals with pollinators and flowering strategies related to them.

It also includes the major events that the various pollen components will be involved in if pollination is successful.

POLLEN DESICCATION AND REHYDRATION (HARMOMEGATHY)

When the tiny pollen grains pass from anther to atmosphere, the protoplast can lose much of its water in just seconds. Heslop-Harrison (1979) reported that losses of 15–35% of fresh weight are known, which really means a large volume of water because most of the weight of a pollen grain is in its relatively massive wall. The vegetative cell is firmly attached to the pollen wall, so when the desiccating protoplast shrinks it pulls the intine inward, which acts in turn on the more rigid exine. Colpi and pores are squeezed shut, spaces between the bacula of the exine are reduced, and the entire grain is often forced into a more concise shape—for example, from spherical to oblate (football-shaped). Pollen desiccation therefore acts like a developing vacuum, applying strong negative pressure on the exine and threatening to cause the entire grain to collapse on itself. Some palynologists feel that the exine serves primarily to withstand these strong internal tensions, and that its other functions are secondary. Bolick (1981) applied engineering principles to the pollen grain wall and demonstrated that it does indeed have structural features similar to those used in construction of buildings to withstand stress.

In two surveys that measured changes in shape and volume of mature pollen after removal from the anther, Payne (1972, 1981) sampled a total of 176 representative species from among dicots and monocots. He found reductions in volume of 4–78%, with a mean value of 46%, which means that exposed pollen grains of many species shrank to half or less of their original volume in the anther. These wrenching internal and external stresses that pollen grains undergo while adjusting to rapid shrinkage from water loss are abruptly reversed when pollen rehydrates on the stigma. A descriptive term for adaptation to these extreme changes in pollen grain volume and shape is harmomegathy ("big harmonic changes"), coined by the early pollen expert R.P. Wodehouse (Payne, 1972). Wodehouse and Payne both felt that harmomegathy has been a major factor influencing the evolution of the pollen wall. Hesse (2000) has discussed in considerable detail how the various pollen wall layers are affected during harmomegathic changes, and Rowley and Skvarla (2000) have demonstrated in a hydraulic press that the exine of many species has considerable ability to adapt to, or withstand, deformation.

LIFE SPAN OF POLLEN

Many seeds, spores, and various vegetative disseminules are able to remain dormant for long periods after desiccation and revive when rehydrated. But pollen grains are inherently shortlived. Bi-celled pollen has a life span of 2 to a few days, while tri-celled pollen is viable for only about 12 hours to 2 days. Although pollen is always short-lived, its natural life span is difficult to determine precisely (Dafni and Firmage, 2000). Because mature pollen is packed with starchy and lipoidal food reserves (see following section), the reason for the short life span is evidently not starvation. An explanation remains to be discovered.

Artificial storage techniques have extended the life span of some pollen. Quick-freezing, for example, has stretched the viability of some commercially important species of bi-celled pollen by several months to two years, and tricelled pollen from several days to as long as a few months, respectively (Bajaj, 1987). It is of considerable practical importance to be able to store pollen for long periods, and to be able to ship samples of desired pollen for breeding purposes, as is done routinely for many kinds of animal sperm. Shivanna (2003) includes a detailed discussion of pollen viability and the techniques used to study and enhance it.

POLLEN FOOD RESERVES

The vegetative cell of a pollen grain has either mostly starch or mostly lipid food reserves, but never 100% of either. Baker and Baker (1979) verified this in a survey of pollen food reserves in about 1,000 species distributed among many dicot and monocot families. They also showed that pollen with a certain type of food reserve tends to be associated with a certain mode of pollination. Starch-rich pollen is associated with self-pollination and pollination by wind, birds, and lepidopterans (moths and butterflies). Lipid-rich pollen is associated instead with bee and fly pollination. Pollen grains smaller than 25 ¼m in diameter are, however, always lipid-rich regardless of their mode of pollination, probably because lipids pack more energy than starch in a smaller volume. In some families all pollen seems to be of one kind; in other families both kinds of pollen are represented.

FACTORS IN POLLINATION SUCCESS OR FAILURE

The probability of successful pollen transfer from anther to stigma depends on the efficiency of the mode of pollination of a species. For example, the pollen of a permanently closed (cleistogamous), self-pollinating flower has a high probability of reaching the stigma, as expected. In self-pollinated open (chasmogamous) flowers, the probability of success is somewhat less. In animal-pollinated species, much depends on how closely the pollinator restricts itself to one or a few species. Wind is the most inefficient cross-pollinator, because pollen is then almost completely dependent on chance. The number of pollen grains produced by flowers of a particular species is correlated in general with its pollination strategy.

The type of food reserve in a pollen grain tends to attract certain kinds of pollinators; in addition, volatile compounds in the pollenkitt on the pollen surface, or absorbed from other floral structures, can emit characteristic and often species-specific odors, which also attracts certain pollinators. Other pollen compounds, volatile or otherwise, have been shown to inhibit insects from eating some pollen grains, and certain pollen compounds have even been shown to have antimicrobial properties. These unexpectedly diverse properties of pollen related to their pollinators have been reviewed comprehensively by Dobson and Bergstrom (2000).

Other factors that affect the success of pollen transfer include structural features of flowers, time of pollen release relative to when the stigma is receptive, amount of pollen produced, and the vagaries of weather. Much of the pollen of many plants is eaten by pollinators—for example, the pollen made into honey by bees. All of these variables contribute to great losses among pollen. It is likely that most pollen grains from most species either never touch a stigma or they land on the wrong stigma.

Wind-borne pollen of some species causes human allergy. Such pollen is in the air in enormous quantities at certain times, and when inhaled by people, they stick to the wet mucous surface of the respiratory tract. This provides a stigma-like environment for many kinds of pollen, which react by releasing their normal recognition proteins and other substances. The proteins released by pollen of some species (ragweed and many others) trigger antibody release by host cells, and a series of reactions follow that result in mild-to-severe allergic reactions in many people. A recent detailed discussion of human/pollen allergy interaction is in Shivanna (2003).

Pollen is often deposited on the wrong stigma. The stigma in such a situation might simply be unresponsive, or it might react by either preventing the pollen from germinating or stopping the pollen tube at some point after germination. Pollen-stigma interactions vary greatly because pollen and stigmas differ considerably in structure and biochemistry. Favorable or unfavorable reactions can occur on the stigma surface, or the interaction may be delayed until the pollen tube is growing in the style, or even in the ovary. All of the possibilities can be accommodated, however, within three categories: incongruity, incompatible interaction, and compatible interaction. Shivanna (2003) provides a detailed account of these topics. The treatment presented here is more general but includes some examples.

POLLEN-STIGMA INTERACTION: INCONGRUITY

Incongruity is really "no reaction" because in this situation the pollen grain and the stigma on which it lands produce such different genetic (i.e., biochemical) messages that they can be said to simply pass by each other undetected. As Knox (1984) has put it, " . . . incongruity is an uncontrolled process arising from the lack of co-adaptation." For example, imagine that some pollen from a wind-pollinated oak tree floats through an open window on a spring day and lands on the receptive stigma of an African violet flower. The oak pollen will exude a set of biochemical messages, and the African violet stigma also has one or more recognition and stimulatory substance(s) on its surface, but these sets of chemicals are too unlike each other to react with each other.

As a result the oak pollen will either not germinate or it might germinate but produce only a short tube that soon stops growing. It will perish in spite of having landed in what seems to be a perfect nurturing location.

Real examples of incongruity were described by Heslop-Harrison (1982) from crosses between distantly related grass species. Pollen behavior ranged from no germination to tube growth that stopped somewhere in the style short of the ovary. No chemical recognition interaction occurred. He speculated that the reasons for this behavior included a difference in osmotic balance between the pollen tube and adjacent cells of the transmitting tissue; or the pollen tube might simply have been of the wrong diameter to be accommodated within the stigma.

The concept of incongruity can perhaps be conveyed by the simple analogy of a visitor to a house who signals the inhabitants by knocking or ringing the doorbell, but those inside do not hear it and therefore do not answer. Those inside make no decision to let the outsider come in or stay out, and the visitor receives no response. The end result is, however, the same as rejection.

POLLEN-STIGMA INTERACTION: INCOMPATIBILITY

Incompatibility might be considered to include incongruity as a subset or special case. But incompatibility is really different, at least in theory, because it requires active rejection based on genetic (biochemical) messages released from pollen similar enough to those of the carpel to be recognized and responded to. True incompatibility therefore appears to be restricted to pollen and stigmas belonging to at least reasonably closely related plants. To continue the visitor analogy from the previous section, incompatibility requires that those inside hear the bell or knock, recognize that there is someone outside, but refuse entry. Interaction of a sort is involved here even though the end result is rejection, as in incongruity.

Both incongruity and incompatibility were demonstrated among various crosses when pollen from species in the same or different families was placed on stigmas of other *Crocus* or *Gladiolus* species (monocots, both Iridaceae). Species of both genera are self-compatible, with exposed stigmas. Pollen from 22 species, 5 from Iridaceae and 17 from other families, were placed on the stigma. Pollen from 10 species failed to become hydrated, perhaps indicating incongruity. Possible incompatible reactions occurred when pollen from another 10 species germinated but did not penetrate the stigma surface, and 2 other species produced pollen tubes that extended into the style before they stopped growing. These results were reported by Knox and Clarke (1980) as part of their review of pollen and carpel interaction.

Varied pollen behavior has also been reported from crosses made within the grass family. Heslop-Harrison (1982) integrated information from the literature with his own studies and reported that, in self-incompatible species, pollen placed on the stigma of the same plant exhibited signs of inhibition and rejection either just after the pollen tube contacted the dry stigma surface or after it penetrated the stigma cuticle and had grown for a short distance between stigma papilla cells. A conspicuous callose deposit in the pollen tube wall where a tube stopped growing indicated that an incompatibile reaction had occurred.

When incompatible crosses were made between grass species, however, callose typically did not form; instead, pollen tubes more commonly swelled and/or burst. The tubes grew for various distances in the stigma or style, or some even continued into the ovary, where their growth appeared directionless and disoriented, missing the micropyle. Pollen tubes of some crosses did reach the embryo sac and effect fertilization, but the hybrid embryo or endosperm eventually aborted. Heslop-Harrison (1982) suggested that the wide range of behavior of grass pollen tubes during these crosses indicates that early-acting (i.e., on the stigma) cross-incompatibility is linked to the self-incompatibility system of recognition and rejection, whereas lateacting (i.e., in style or ovary) incompatibility is incongruity. Similar pollen tube behavior in self-incompatible legumes has also been documented among species of *Medicago* (Sangduen et al., 1983) and *Trifolium* (White and Williams, 1976; Williams and White, 1976).

Among the few detailed studies of interspecific incompatibility crosses, that of Williams et al. (1982) on some Australian *Rhododendron* species (dicot family Ericaceae) merits mention. These commonly grown shrubs and small trees have flowers with a wet stigma and a hollow style. Foreign pollen was taken either from different species of *Rhododendron* or from various species of related genera within the Ericaceae and crossed to a single *Rhododendron* species.

Pollen from the same cross always behaved in the same way: either it always remained inert on the stigma; or the pollen tube always stopped at the same place in the carpel, where its tip enlarged, coiled, twisted, burst, or showed unusual patterns of callose deposition. Williams et al. (1982) concluded from these crosses that there was little evidence for selfincompatible interaction. Instead, they felt that "The situation in *Rhododendron* appears to represent a more generalized or diverse control over pollen tube growth, expressed at a number of different sites in the pistil [gynoecium] as a spectrum of errors in macromolecular carbohydrate metabolism." In other words. incongruity was implicated, because each different pollen type sends out its own genetic message, the reactions leading to blocked tube growth are diverse, and they occur at different places within the gynoecium.

POLLEN-STIGMA INTERACTION: SELF-INCOMPATIBILITY

If a species exhibits cross-incompatibility, the carpel will reject foreign pollen that exhibit small-to-moderate differences from its own recognition factors. Self-incompatibility (SI), however, is the rejection of self, which means that identical recognition signals will prevent pollen germination or stop pollen tube growth. In species with SI, pollen that lands on the stigma of its own flower, or on the stigma of other flowers on the same plant, will be recognized but then rejected. The rejection may occur even before a pollen grain germinates, or the pollen tube may be stopped somewhere between the stigma and the ovary.

Self-rejection in sexual reproduction might seem paradoxical when compared to a vegetative graft union, in which success is dependent on self-acceptance. The two must, however, be quite different phenomena. SI in sexual reproduction prevents inbreeding, which promotes genetic diversity. SI is widespread, having been reported from at least 100 families of both dicots and monocots (Gaude and Dumas, 1987).

One might expect that it should be even more common, since SI is a mechanism that promotes genetic diversity. But here again one must consider ecological complexities and pollination strategies that could favor self-compatibility. The two main categories of SI are heteromorphy and homomorphy. Heteromorphy ("differing forms") occurs when a single species has two or three morphologically distinct flowers.

These flowers differ from each other in their style length, which is the main character used to distinguish the forms: distyly means two forms (short- and long-styled) and tristyly means three forms (short-, mid-, and long-styled). Heteromorphy is, in fact, often referred to as heterostyly. Stamen lengths and pollen and stigma size and shape may also differ from each other. Successful pollinations can occur only between two different floral forms, not between identical forms.

Heteromorphy occurs in at least some members of 24 scattered families, mostly dicots, and it seems to have evolved independently several times. Buckwheat (*Fagopyrum esculentum*, dicot family Polygonaceae) is probably the only crop plant exhibiting heteromorphy, but *Forsythia* (Oleaceae) and primrose (*Primula*, Primulaceae) are two other familiar cultivated genera with heteromorphic species. Heteromorphic SI will not be described here any further, although it has intrigued botanists at least since Charles Darwin's 19th century studies. Richards (1986) provides a thorough discussion of heteromorphy, and Barrett et al. (2000) is an entry into more recent studies.

Homomorphy ("one form") is far more common than heteromorphy, and it involves species with only one form of flower. Almost 100 families have some representatives with homomorphic SI, and continuing investigation among tropical species indicates that many more of them also have SI. Richards (1986) estimated that possibly more than 50% of angiosperms exhibit self-incompatibility, mostly of the "gametophytic" type, which will be described later in this chapter.

Homomorphy has been investigated extensively since the early 20th century; studies have steadily accumulated, and the subject is important enough to be periodically reviewed—e.g., Barrett (1988), Gaude and Dumas (1987), Nettancourt (1984), Richards (1986), and Seavey and Bawa (1986). Only the general features of this phenomenon, and some examples, will be presented here.

Homomorphic SI most commonly acts through a single gene with several possible alternative expressions, or alleles. In the mustard family (Brassicaceae), for example, from 24–50 of these "S-alleles" have been described in various cultivated members (Gaude and Dumas, 1987). A more complex SI mechanism is reported from grasses, which have two genes, each multiallelic (Baumann et al., 2000), and in sugar beet, which has four multiallelic genes.

The basic idea in homomorphy is that if one or more S-alleles (depending on the species) residing in the pollen are identical to those that occur somewhere in the carpel, usually in the stigma or style, pollen will either not germinate or it will germinate but pollen tube growth will be arrested at some intermediate level. Why should two identical genetic messages cancel each other out and thereby prevent normal pollen tube growth? The cessation of pollen growth in SI could be viewed as either of the following:

- Oppositional—i.e., an active inhibition by the carpel, perhaps from an unfavorable chemical reaction or series of reactions
- Complementary—i.e., incompatible pollen grains are simply unable to stimulate the carpel to transfer enough, or even any, of the critical substances needed to sustain pollen tube growth all the way to the embryo sac.

Vasil (1987) has pointed out that pollen tubes of very few species will grow to the length needed to reach the ovules when germinated on an artificial medium; most will grow to only a tenth or less of the required distance. This indicates that pollen tubes need to have either nutrients or growth substances, or both, provided by the carpel.

The two contrasting oppositional and complementary hypotheses of self-incompatibility have also been invoked to explain, or at least to initiate speculation in, many other biological situations as well (Gaude and Dumas, 1987). These hypotheses are simply two logical scenarios to consider in the absence of firm evidence. Baumann et al. (2000) admit that an explanation of SI at the molecular level is unclear, although they feel that proteins are implicated.

Two subdivisions of homomorphic SI are recognized. If there is only one allele, which has been produced entirely by the haploid genome of the male gametophyte (pollen grain), the species belongs in the category called gametophytic-SI or G-SI. This is the most common type of SI (Richards, 1986). Plants with G-SI have bi-celled pollen, with the products of the S-allele located either in the cytoplasm or embedded in the intine of the pollen wall.

The stigma is of the wet type. A common reaction in G-SI is callose deposition in the pollen tube, typically appearing after the tube has grown into the style, which stops further tube growth. The haploid pollen can express only one possible S-allele, whereas the diploid carpel has two. Incompatibility results if the single pollen S-allele matches either of the two present in the carpel.

An example of G-SI was described by Kahn and DeMason (1986) in their study of pollen tube growth and fertilization in certain commercial *Citrus* species and varieties. They found that when the self-incompatible Orlando tangelo was self-pollinated, most pollen tubes stopped growing while still on the wet stigma, a few penetrated the upper part of the style, and at least one grew into the lower part of the style. The arrested tubes showed their incompatibility by several possible abnormalities: heavy callose deposition at the tip, irregular callose deposition in the tube wall, a spiral pattern of tube growth, and tapered or burst tips. These reactions from just one self-incompatible pollination match the range of growth anomalies described earlier from cross-pollinations in *Rhododendron* (Williams et al., 1982). Another example of G-SI is sweet cherry, *Prunus avium* (Rosaceae). Raff et al. (1981) described the receptive cherry stigma as covered with a copious secretion containing proteins, carbohydrates, and lipids. Selfpollination resulted in pollen tubes penetrating the solid transmitting tissue of the upper style, where the tubes stopped growing and usually exhibited swollen tips with callose. In crosspollinations, in contrast, pollen tubes grew normally and fertilization was successful.

What is perhaps surprising is that pollen of two other *Prunus* species germinated on the stigma of sweet cherry, and their tubes grew normally all the way into the ovary (but fertilization was not tested). Pollen from certain other genera of Rosaceae were also placed on the sweet cherry stigma; these grains germinated but the tubes stopped growing in the upper part of the style. Pollen from species of other families did not germinate on the sweet cherry stigma. Sweet cherry flowers evidently have a strong aversion to self-pollen, and either cross-incompatibility or incongruity occurs from wide crosses. The flowers seemingly cannot distinguish their own pollen from pollen of other individuals and other species of *Prunus*, or else the incompatibility occurs in the ovary, which results in sterility since only one ovule is produced.

In a second type of homomorphic SI the Salleles are not produced within the male gametophyte itself but are instead expressed in the form of proteins produced by the tapetum of the parent sporophyte and deposited on the exine of the pollen wall either by active tapetal secretion or by tapetal cytoplasm as a result of the rupture of tapetal cells. This homomorphic category is called "sporophytic-SI (S-SI). The pollen carries two sets of S-alleles on its surface because they came from the sporophyte (tapetum), which has two sets of chromosomes; therefore, the reaction on the stigma is more complicated. If just one allele is matched, but it happens to be a dominant or co-dominant allele, incompatibility occurs. If the single allele that is matched is recessive, the pollen may overcome the match and produce a compatible fertilization. S-SI seems to be a derived condition, according to Richards (1986), because it occurs in just a few of the more advanced families—e.g., Brassicaceae and Solanaceae. At the molecular level several interacting proteins have been identified, although some key pieces of the puzzle are missing, according to McCubbin and Kao (2000).

The S-SI reaction is often reported to stimulate callose deposits in both the pollen tube and in stigma cells in contact with the tip of the tube. This mutual response has been viewed as a visible sign of pollen tube arrest. Unlike the callose deposition that occurs during microsporogenesis and megasporogenesis, which is a part of normal development, callose that forms in response to incompatibility seems to be more of a rapid reaction to disturbance, similar to the quickly formed callose that may appear at the surface of living plant cells exposed by wounding. The ability of the stigma and/or style to elicit the callose response is defensive, a reaction that one can speculate is necessary to maintain the genetic integrity of the species.

THE MENTOR POLLEN TECHNIQUE

The reasons advanced to explain both G- and S-incompatibility on the stigma and in the style are supported, in a practical way, by the use of mentor pollen, compatible pollen that is placed on a stigma along with incompatible pollen. The simplest procedure is to mix some living compatible mentor pollen in a certain proportion with some living incompatible pollen, and then place them together on a stigma. The compatible pollen acts as a mentor (a guide) by releasing enough favorable chemical recognition signals to either dilute or mask in some way the incompatible chemical message exuded by the incompatible pollen. This fools the stigma or style (or both) into accepting the incompatible pollen tubes along with the compatible ones, thus allowing an otherwise impossible fertilization to occur.

Successful germination is not enough, however, because tubes emerging from the incompatible pollen must grow fast enough to reach at least some of the ovules ahead of the compatible pollen tubes, and the latter usually grow faster. Partly to overcome this problem, a more sophisticated method uses dead mentor pollen killed by radiation, heat, or some other method that is gentle enough to preserve the recognition proteins. Mixed with living incompatible pollen, the dead compatible mentor grains can still exude activating recognition molecules but will not germinate and compete with the desirable incompatible pollen tubes.

Mentor pollen techniques do not always work, but enough successful demonstrations have been shown that it must be considered useful both for plant breeding and to help understand incompatible reactions. Detailed reviews of this subject can be found in Stetler and Ager (1984) and Knox et al. (1987), and a recent concise summary appears in Shivanna (2003).

CALLOSE AND INCOMPATIBILITY

Callose is common in a plant, as discussed in other chapters of this book. It is a highly selective filter, or even a complete barrier, to the passage of molecules. In both S-SI and G-SI incompatibility, however, callose can be viewed instead as a physical barrier that forms quickly and stops further pollen tube growth. In *Brassica*, which has S-SI, Kerhoas et al. (1983) removed tryphine of tapetal origin from the surface of pollen grains and placed it on both incompatible and compatible stigmas. A tiny dollop of tryphine by itself stimulated callose to form only on incompatible stigmas. Since tryphine is a pollen wall coating contributed by the tapetum, which is part of the sporophyte, this demonstrated that sporophytic proteins affect pollen germination in this species.

Stott (1972) showed that incompatible pollen tubes in several apple cultivars grew very slowly through the style and deposited a continuous thin callose layer inside the tube wall (callose plugs might also have formed at intervals across the tube, but they were obscured by the lateral wall deposits). Compatible pollen tubes, in contrast, grew much faster and were dotted only intermittently with transverse callose plugs. When a mixture of compatible and incompatible pollen was germinated on one stigma, there was evidently no mentor effect, because each tube behaved according to its compatible or incompatible nature as just described.

The incompatible reaction here was not complete because in a real situation the faster-growing compatible tubes would have reached the ovules first and fertilized them while the incompatible tubes would simply degenerate somewhere else in the carpel.

LATE-ACTING (OVARIAN) SELFINCOMPATIBILITY

Studies on self-incompatibility have concentrated on species in which rejection occurs on the stigma or in the style, either because this is a more typical situation or because it is easier to conduct experiments when reactions occur at these more easily detectable sites. But there are also many published examples of pollen tubes entering the ovary, and even the ovule, before self-rejection occurs. Seavey and Bawa (1986) reviewed these published examples, which they called "late-acting self-incompatibility."

The details of how and why rejection should sometimes be delayed until the pollen tubes reach the ovary are too complex and too tentatively explained as yet to be dealt with here in any detail. Seavey and Bawa (1986) felt that it might sometimes be of advantage to the ovulebearing parent plant to allow pollen tubes to compete with each other by growing for a relatively long distance, and in this way to be "evaluated" after they have passed through the stigma and style. A rejection reaction that occurs before a pollen tube enters an ovule requires no new SI mechanism to be postulated. But rejection within an ovule seems to call for a new kind of reaction that is distinct from either G-SI or S-SI. Seavey and Bawa regarded the destruction of a small percentage of the total ovules by degeneration of self-fertilized embryos and/or endosperm as a negligible loss for most plants.

MOLECULAR BASIS FOR POLLENSTIGMA INTERACTIONS

Numerous studies have been conducted to discover and explain the molecular interactions that control recognition and rejection of pollen. In dry stigmas the pellicle is involved and lectins (glycoproteins with the ability to bind to specific sugars) have been implicated. According to Gaude and Dumas (1987), " . . . the chemistry of the recognition event, which may be defined as the interaction between products of the S gene and the metabolism modifications that follow, are still unknown." Clarke et al. (1985) reviewed the progress of their research on the molecular basis for self-incompatibility in tobacco and were also puzzled as to how identical S-allele products could interact to inhibit pollen growth. They suggested the possibility of post-transcriptional changes, which means that what starts as identical chemical pathways would not end up with identical products. This remains to be demonstrated.

The molecular interactions that control selfincompatibility have been studied most in the mustard family (Brassicaceae), and some examples were cited earlier in this chapter. Lord and Russell (2002) reviewed many of the investigations, which have revealed a host of possible involved compounds, especially proteins, but none of them can be said to conclusively explain self-incompatibility. Promising clues have been uncovered, but they remain tentative because they " . . . suggest a complex series of events in both adhesion and hydration of pollen on dry stigmas."

COMPATIBLE INTERACTION

This chapter began with an examination of those features of pollen that enable the grains to withstand desiccation and the threat of collapse and a description of the nature of pollen food reserves in relation to types of pollinators. The possible ways that pollen and pollen tubes can be inhibited or stopped in the carpel were then considered by describing various incongruous and incompatible reactions. Since most pollen that is dispersed naturally is unsuccessful— that is, either reaches no stigma or lands on the wrong stigma—this seemed appropriate. Successful compatible pollination will now be taken up.

But even if the right pollen grain lands on a compatible stigma there is a further possible barrier because the receiving stigma must be in a receptive condition if the grain is to germinate. Just as pollen grains have a certain short life span, so does the receptive period of a stigma. Heslop-Harrison (2000) viewed stigmas as "control gates" and discussed their life span in various species, which might be as short as one hour in oats (*Avena*) or as long as seven days in some species. The stigma of most species, however, is receptive for a time period somewhere between these extremes.

Receptive stigmas are of two types: wet or dry. In a wet stigma, pollen simply sinks into the viscous fluid secreted onto the surface. The time required for such a grain to rehydrate and germinate depends partly on the proportion of water to other components in the stigmatic secretion, coupled with the inherent properties of the pollen grain. Stigmatic secretions are composed of various proportions of some or all of these substances: lipids, amino acids, peptides, proteins, polysaccharides, phenolics, and water. The composition varies, of course, depending on the species (Linskens and Kroh, 1970).

Some wet stigmas become wet only after being stimulated by contact with pollen, a behavior that occurs in scattered unrelated plants. The stigma of watermelon (*Citrullus lanatus*, dicot family Cucurbitaceae), for example, begins to secrete copiously about 10 minutes after pollen lands on it, and foreign pollen as well as compatible pollen elicits the secretion. Glass beads substituted for pollen do not do so, however, indicating that the secretion response does not simply result from physical contact (Sedgley, 1981). In sugar apple (*Annona squamosa,* dicot family Annonaceae), a subtropical fruit species, copious secretion does not cover the stigma until after about 40 minutes following pollination (Vithanage, 1984).

A dry stigma never has a copious surface secretion, but it has three possibilities for pollen attraction and retention: electrostatic force, surface tension, and adherence by chemical bonds. Electrostatic force means attraction through static electricity, which seems reasonable to expect if a dry pollen grain strikes a dry stigma. Since most pollination occurs during dry weather, an electrostatic force could allow a dry stigma to attract and hold pollen briefly, perhaps for only about a second, until permanent attachment is secured by some rapidly secreted organic substance.

Wottiez and Willemse (1979) experimented with both natural dry stigmas and artificial metal ones. They concluded that electrostatic attraction occurs but that it is limited in its significance. Experiments conducted by Corbet et al. (1982) on the ability of rapeseed pollen to jump between two charged pins, and from anthers to honeybees and from there to stigmas showed that pollen could be attracted over a distance of up to about half a millimeter. Although a miniscule distance, both wind- and insect-pollination could involve such electrostatic attraction. Challoner (1986) speculated that one of the functions of the various sculptured features found on the pollen exine might be to slow down dissipation of the electrical charge from the pollen grain, thus maintaining electrostatic attraction for a longer time. A recent review by Vaknin et al. (2000) provided details from over 40 publications concerning this subject; they concluded that electrostatic phenomena are real and important both for transporting pollen (e.g., adhering it to insect body hairs) and for brief retention of pollen by dry stigmas.

A few studies have demonstrated that a chemical bond forms where the pollen exine contacts the pellicle of a dry stigma, or where some of the tapetal coating on the exine surface of some grains flows onto the stigma surface. This chemical bonding, which is a kind of glue, has been reported to adhere pollen to the stigma. Ferrari et al. (1985), for example, regarded this as one of their four stages of pollen capture in mustard, *Brassica oleracea*. Three proteins reported from the stigma, but none from the pollen, have also been implicated in the adhesion of pollen in this species (Dickinson and Roberts, 1986). In grasses, Heslop-Harrison and Heslop-Harrison (1985) mentioned a similar kind of bonding. The inability to establish such a bond, or the formation of only a weak one, has been shown to be the initial event associated with rejection of incompatible pollen in some species. Ferrari et al. (1985) were able to remove incompatible pollen of mustard easily from the stigma, but compatible pollen remained firmly attached. But a recent review of pollen-to-stigma adhesion among mustards (Heitzmann et al., 2000) concluded "In spite of years of investigation by many laboratories, the precise mechanism by which two cells specifically adhere to one another is not known."

The pollen grain is a tiny spheroid that quickly becomes dehydrated and shrunken upon exposure to air. Very shortly after it has established at least a temporary contact with the stigma, water begins to flow back into it. Heslop-Harrison (1979) reported that this initial influx is conspicuous in grass pollen because the vegetative cell membrane, which normally controls molecular movements, cannot regulate water movement in its dehydrated state; thus water flows into the grain at an uncontrolled rate. The excess water exudes from the grain until the cell membrane reestablishes its normal organization and is able to control osmosis.

Heslop- Harrison showed this phenomenon in rye (*Secale cereale*) using a series of photographs taken at short intervals during just 80 seconds (Fig. 7.2). A meniscus is formed by the water and dissolved pellicle components between grain and stigma, which provides an additional force that helps bind the rye pollen to the stigma. In a mustard (*Brassica*) flower, pollen behaves much differently, rehydrating very slowly after adhesion to the stigma, which requires 1–4 hours. Dickinson and Roberts (1986) felt that the tryphine coating on the mustard pollen exine may resist the flow of water into the pollen cytoplasm. *Brassica* has been studied intensively with respect to compatible and self-incompatible reactions on the stigma. A diagram of pollen and stigma behavior in both situations provides a summary for much of this chapter with respect to possible interactions on the stigma.

A compatible pollen grain either sinks into a wet stigma's secretions or becomes attached firmly to a dry stigma by some combination of electrostatic forces, surface tension, and/or chemical bonding. The pollen protoplast then becomes rehydrated and its metabolism increases to some peak level. Hydrostatic pressure coupled with enzymatic weakening of pores or colpi forces the grain to germinate and form a tube that will protrude from one pore or colpus. Pollen germination, and the subsequent events that lead to fertilization, are subjects for the next chapter.

8

Pollen Germination, Pollen Tube Growth, and Double Fertilization

The hazardous journey of a pollen grain from anther to stigma, and the various possibilities for acceptance or rejection after a grain arrives on a stigma, were subjects for Chapter 7. This chapter takes up events after a compatible pollen grain has satisfied the stigma gatekeeper. The pollen protoplast squeezes itself out through an aperture in the exine, and traverses the stigma, style, and ovary within its progressively elongating flexible tube until it finally penetrates a dying cell in the ovule and bursts, expelling the sperm cells into foreign protoplasm where they will seek out two different cells and merge with their respective nuclei. This scenario seems highly imaginative, and so it is not surprising that pieces of the puzzle were not all fitted into a coherent account until late in the 19th century. This chapter takes up these unlikely events that end with double fertilization.

GERMINATION AND EARLY TUBE GROWTH

The pollen grain that lands on a compatible stigma is either bi-celled (the most common type) or tri-celled. Pollen of both types arrive desiccated and must absorb water from the stigma to become rehydrated and turgid again. Their internal membranes can reconstitute themselves, which in turn allows normal water relations to be reasserted to allow metabolism to rise.

Pollen respiration is close to zero in the dehydrated state; it increases with rehydration, but not equally fast in all pollen.

Tri-celled pollen is reported to respire 3–4 times faster than bi-celled pollen, but for a shorter period of time. This generalization is, unfortunately, only supported by a tiny sample of species (Hoekstra and Bruinsma, 1978; Hoekstra, 1979). The respiration difference might be related to the presence of fully formed mitochondria in tricelled pollen as compared to only immature mitochondria in bi-celled pollen, although this has been determined for only one tri-celled species of *Aster* (Hoekstra, 1979). The bicelled pollen that Hoekstra studied showed various states of mitochondrial maturation after rehydration, but all had slower respiration rates than *Aster*. These mitochondrial and respirational differences show up finally in germination, which for tri-celled pollen takes only a few to several minutes, as in grass pollen; bicelled pollen may take 1–2 hours or even longer (Hoekstra and Bruinsma, 1978).

Tri-celled pollen, is viable for just a few hours to about two days. Such pollen requires fast pollination followed by rapid germination. Bi- and tri-celled pollen tubes also have different rates of growth, a topic taken up later in this chapter. One can speculate that a temporary slowing of tube growth in bi-celled pollen occurs because the generative cell divides to form sperm cells, thereby possibly affecting the overall pollen tube growth rate. Forming sperm cells while still in the developing pollen grain could require functional mitochondria in tri-celled pollen, and it could also use up enough pollen food resources to partly explain the shorter life span of such pollen.

The first visible sign of germination in a rehydrated and turgid pollen grain is intine protruding from an aperture. If a pollen grain has more than one aperture it is reasonable to wonder whether the tube emerges simply by chance through any one of them, or does it exit through the one that is closest to the stigma surface? On a dry stigma the latter seems reasonable, but for pollen submerged in stigmatic fluid all apertures appear to offer the same opportunity. There are reports and illustrations of tubes emerging into the air and then turning around and growing back into the stigma, a possible indication that the pollen tube has no control over which aperture it will emerge through. Little seems to be known about what, if anything, determines the exit aperture in grains with more than one aperture.

The intine of a pollen grain has three layers: outer pectin-rich, middle protein-containing, and inner cellulosic-callosic. The intine is typically thicker under the apertures; elsewhere, the outer and middle intine layers are thinner and may even be absent. If a thin exine layer is present over the aperture, some reports describe it as exposing the intine by opening like a ship's hinged porthole. As the intine bulges beyond the aperture, its outer layer becomes hydrated and is soon dispersed in the stigmatic fluid. The middle intine layer also dissolves away but as it disperses it releases packets of proteins, which include enzymes; these intine components will become active in pollen/stigma recognition reactions and cuticle digestion, and they may also participate in other enzymatic processes that must occur during early pollen tube growth and penetration of the stigma. While the two outer intine layers dissolve, the inner intine layer is retained and will become the pollen tube wall; new wall material will be added continuously to it near the tip of the elongating tube.

These germination events have been described in the cereal grass, rye (*Secale cereale*), by Heslop-Harrison (1979, 1987). Ungerminated rye pollen shows the thick, pectic oncus (lump or growth) or Zwischenkorper ("interspersed body") The hydrated outer pectic intine layer, now turned into a dispersed gel that pushes the operculum out before it. The underlying emerging pollen tube tip can be seen, which at this stage has a wall composed of the protruding inner intine layer. Continued rye pollen tube growth on a semisolid gel medium shows a translucent apical zone free of organelles; this zone is filled instead with polysaccharide vesicles (not detectable at this magnification) that will provide new wall material. Proteins secreted from the tip are visible in this figure as a thin, darkstaining cap.

At a somewhat later stage the apical clear zone becomes larger, and proteins are secreted largely from the flanks of the tube. These secreted extracellular proteins are necessary for continued chemical interaction with the transmitting tissue of the carpel; they maintain compatibility and also soften, digest, and absorb for nutrition the intercellular material through which the tube grows toward the ovary. The germination story described for rye has been verified for the lily pollen tube by Miki- Hirosige and Nakamura (1982). A detailed and rather dramatic drawing from their paper shows polysaccharide particles crowded into the tip zone, which may physically exclude the organelles. The three original intine layers extend only to about the ruptured exine margin, beyond which a new layer continues as the thin pollen tube wall; this differs in detail from rye, where it is the inner intine layer that becomes the pollen tube wall. Perhaps different species generate the pollen tube wall in somewhat different ways.

The pollen tube wall grows at the apex, where polysaccharide particles merge with, and add to, the existing loosely arranged microfibrillar wall. The newly formed wall organizes itself just behind the apex into an attenuated version of the 3-layered intine of the original pollen wall, with an outer layer rich in pectins, a middle cellulosic layer, and an inner layer of microfibrils mixed with callose. Heslop-Harrison (1987) estimated that, in a fast-growing grass pollen tube, the three strata must continue to form each 10–13 seconds.

The vegetative nucleus and generative cell have not yet emerged from the pollen grain. This figure lacks microtubules and microfilaments, however, which are two forms of structural cytoplasmic strands that other studies have shown are important in shaping the tube and guiding the movement of the various cytoplasmic components. Pollen tube growth recorded by video camera through the microscope shows regular pathways along which organelles move toward the tip along the periphery of the tube and return via the interior. Heslop-Harrison and Heslop- Harrison (1994) reviewed such studies and speculated that as vesicles containing wall material travel toward the tip, random motion carries some of them to the periphery, where they discharge their contents and add to the wall.

The middle cellulosic layer helps maintain the tubular shape, and the inner layer captures polysaccharide vesicles, which become incorporated into the outer two layers. Vesicles that do not contact the inner wall retain their contents and are recycled to return later to the tip. This cyclic flow is real, but they were puzzled as to how the pollen protoplast can continue moving forward by "crawling" along the older pollen wall, both moving and maintaining its shape at the same time. Recent findings, however, have revealed that the lily pollen tube can obtain traction against a slippery substrate and grow forward in the right direction because a peptide (stigma/stylar cysteine-rich adhesin) combines with a pectic polysaccharide in the style, which adheres the pollen tube to the transmitting tissue (Lord and Russell, 2002).

Microtubules appear to be less involved in pollen tube growth because, when they are prevented from forming by certain chemical inhibitors, neither pollen germination nor shape and growth rate of the pollen tube is affected (Heslop-Harrison, 1987). Microtubules do appear to be involved in stretching the generative cell to fit the narrow pollen tube, and they are present in the tube itself (Heslop- Harrison and Heslop-Harrison, 1988). Another function of microtubules could be to direct cytoplasmic streaming, a process which is necessary, among other functions, to move polysaccharide particles to the tube apex for incorporation into the pollen wall (Heslop-Harrison, 1987). F-actin strands have been reported to guide cytoplasmic streaming in some tube tips (Lord and Russell, 2002).

Microfilaments have been shown by Pierson (1988) to form an irregular weblike network around the periphery of ungerminated pollen grains. After germination the microfilaments converge on the emerging tube and extend into it as a peripheral network. A pollen tube needs strong support to maintain its tubular shape, especially when subject to lateral pressure while growing between cells in the style.

A pollen grain germinating on a compatible stigma must react to chemical recognition molecules, and on a dry stigma the protruding tip of the emerging tube must penetrate the pellicle and cuticle. On both dry and wet stigmas with solid styles, the emerging pollen tube tip must also weaken the middle lamella between stigma epidermal cells in order to penetrate beneath and reach the transmitting tissue. The tip first secretes cutinase to dissolve a hole through the cuticle, and several studies provide direct or indirect evidence for this, as Heslop- Harrison (1987) reported. Some investigators have even been able to cause cuticular digestion just by collecting such pollen tube secretions and applying them alone to the stigma. This is, of course, just one function of the pollen tube secretion, which must also involve other reactions to maintain compatibility and nutritional relations with the carpel.

Recent studies of molecular events on the stigma show that in various species different substances act to guide newly germinated pollen tubes to the transmitting tissue (Lord and Russell, 2002). Lipids in some members of Solanaceae cause a water gradient that attracts the tube. In the wet stigma of lily, which opens directly into the open style, the SCA peptide and pectic polysaccharide mentioned earlier provide adhesion and guidance for pollen tubes.

How long does it take for pollen to show visible signs of germination after it lands on a stigma? A sampling of species shows that there is an overlapping of the ranges for the two pollen types, although tri-celled pollen is generally faster. Germination time for pollen in general ranges from about 1 minute to about an hour, and a few longer germination times are also known. The great variation in pollen structure and tapetal coatings, stigma structure and the nature of the recognition substances that must react upon contact by pollen, as well as the genetic variables of compatibility, offer such a spectrum of possible factors that it should not be surprising that germination times can differ greatly. Some examples are given in the following paragraphs.

Heslop-Harrison (1979) reviewed published reports for tri-celled pollen of several cereal grasses and found a range of germination times from somewhat less than 1 minute to about 10 minutes; for other families, Hoekstra and Bruinsma (1978) reported only 3 minutes for *Aster tripolium* (dicot family Asteraceae), but spinach (dicot family Chenopodiaceae) pollen requires 10–20 minutes (Wilms, 1980). For bi-celled pollen, Hoekstra and Bruinsma (1978) reported germination times of

6 minutes (*Tradescantia paludosa*, monocot family Commelinaceae), 35 minutes (*Nicotiana alata*), and 70 minutes for common cattail (*Typha latifolia*, monocot family Typhaceae). Peanut (*Arachis hypogea*) takes only 5–10 minutes (Bhatnagar et al., 1973), Peruvian tomato (*Lycopersicon peruvianum*) takes 45 minutes (Cresti et al., 1977), and gladiolus (*Gladiolus* sp., monocot family Iridaceae) requires about one hour to germinate (Ameele, 1982).

CELLS AND NUCLEI WITHIN THE POLLEN TUBE

As the emerging pollen tube begins to elongate, for a short period of time the vegetative cell protoplast continues to fill both the shell of the grain and the tube. But soon the entire protoplast (vegetative cell and included generative cell or two sperm cells) flows into the apical region of the elongating tube, leaving the grain and a gradually lengthening empty adjacent portion of the tube itself behind. For an analogy to pollen tube growth consider a skyscraper in which an imaginary elevator forms its own shaft, starting at the top floor, and descends by resting on its advancing front. Instead of leaving a completely empty shaft behind itself, however, as the car descends it forms a horizontal barrier across the shaft every few floors. The elevator is therefore a capsule moving in a vertical shaft of its own manufacture, in which the already traversed portion becomes repeatedly subdivided into empty compartments as the car passes downward. The only vital area is where the elevator car happens to be at any particular moment.

As the vegetative cell flows into the tube its nucleus appears to be viable, but it has often been reported to stain more faintly than generative or sperm cell nuclei (Maheshwari, 1950). Steffen (1963a) discussed in considerable detail the characteristics of the vegetative cell and nucleus and the several hypotheses about its function before and just after germination. Many have thought that the vegetative cell nucleus ceases to have any appreciable function shortly after the generative cell forms. One indication that it could still have a function, at least in some species, is its physical connection to the sperm cells in the male germ unit of some species (see next section).

When a bi-celled pollen grain germinates, its vegetative cell protoplast forms the tube while the vegetative nucleus and generative cell flow into it, although not always in a set order. Peruvian tomato provides an example of a tube in which the generative cell nucleus, already in early prophase, is behind the vegetative nucleus. Cooper (1936) showed that even though the generative cell is severely constricted in the narrow lily pollen tube growing in the stylar canal it nevertheless undergoes a normal mitosis. The generative cell is remarkably flexible in accommodating its mitotic apparatus and cell plate to the constriction imposed by the pollen tube. Sperm cells of tri-celled pollen form in the grain before the anther dehiscences, as in spinach, and are transported to the stigma. After germination the sperm cells flow into the tube with the vegetative cell protoplast, where they follow the vegetative nucleus.

Changes in shape and movement among the entities within pollen tubes have been observed in a casual and qualitative way for a long time. Heslop-Harrison and Heslop-Harrison (1987) sorted out the independent movements of the sperm cells from those of the smaller organelles by filming living rye pollen tubes under a microscope with a video camera. They showed one sperm cell overtaking the second one during a time period of about 65 seconds. They also observed sperm cells moving backward as well as forward, and other organelles, such as mitochondria, moving independently of the sperm cells. They interpreted such independent movements as resulting from F-actin filaments of the pollen tube that provided separate lanes within the tube.

DIMORPHIC SPERM CELLS AND THE MALE GERM UNIT

Sperm cells have been described and/or illustrated as identical to each other in perhaps as many as 99% of the studies that include such matters. A few reports of dimorphic sperm cells (two different sizes and/or shapes) were mentioned by Maheshwari (1950), but he regarded them with considerable doubt. Until recent decades the accepted view has been that the sperm cells are identical, independent, and interchangeable.

In a recent survey of 115 species of 56 families using a squash technique coupled with light microscopy, Saito et al. (2002) reported that only five species from five families (about 4%) had dimorphic sperm cells. They concluded that such sperm cell pairs are uncommon but scattered widely among angiosperm families.

Electron microscopy has been used in only a small number of detailed studies of germinating pollen, pollen tubes, or fertilization, but some of these investigations have provided convincing descriptions of physical connections between sperm cells. Connections between the two sperm cells and the vegetative cell nucleus have also been described. If these three become linked as a trio, either before germination or in the pollen tube, the resulting entity is called a "male germ unit," a term coined by Dumas et al. (1985). The existence of male germ units is one step beyond dimorphic sperm cells, and strengthens the hypothesis that sperm cells are not always identical and therefore not always interchangeable.

Double fertilization in such species may occur by sperm cells whose destination (egg cell or central cell) is predetermined by their structure or arrangement in the tube rather than by chance. The first detailed description of physically connected dimorphic sperm cells, by Russell and Cass (1981) and Russell (1984), was from leadwort, *Plumbago* (dicot family Plumbaginaceae), an ornamental shrub in which the embryo sac lacks synergids but has tri-celled pollen. Russell's reconstruction shows the two sperm cells, but with the vegetative nucleus omitted for clarity.

One sperm cell is obviously smaller and shaped differently than the other. If the vegetative nucleus had been included in the larger sperm cell would have been wrapped around it. The larger sperm cell has virtually all of the mitochondria but almost none of the plastids, and the smaller sperm cell has very few mitochondria and virtually all of the plastids. The significance of these mirror image differences, and the possibility of a predetermined target nucleus for each of these sperm cells, has yet to be determined.

Male germ units in tri-celled pollen have been described more recently from a few additonal species, e.g. spinach and two *Brassica* species by McConchie et al. (1985). In the latter study the rapeseed (*Brassica campestris*) male germ unit was reconstructed beautifully from computer-assisted information compiled from seven serially sectioned male germ units. The sperm cells in rapeseed are not Chapter 8: Pollen Germination and Double Fertilization 127 wrapped around the vegetative nucleus, but one sperm is connected to it by a tail about 10 ¼m long that is stiffened by microtubules. The two interconnected sperm cells, unlike those of *Plumbago*, are similar in size and both have mitochondria but no plastids.

Grasses also have tri-celled pollen, and some studies have attempted to determine if dimorphic sperm cells or a male germ unit occurs among cereal grasses. Mogensen and Wagner (1987) found that barley sperm cells do not form a male germ unit until after pollination, just before the sperm cells and vegetative nucleus enter the pollen tube. In mature maize pollen McConchie et al. (1987) found dimorphic sperm cells, one long and slender with a hook at one end, the other shorter; they were not connected to each other or to the vegetative nucleus. Post-pollination behavior was not studied, so a possible later linking up, as in barley, remains unknown. McConchie et al. (1987) also found clusters of mitochondria, some of them unusually elongate, associated with the convoluted and fenestrated surface of the sperm cells. They suggested that these mitochondrial clusters could provide the additional energy that is probably required because of the unusually long distance from silk (stigma) to embryo sac.

Only two detailed studies of dimorphic sperms have included species with bi-celled pollen, one on a dicot, rhododendron (Kaul et al., 1987), and one on a monocot, amaryllis (Mogensen, 1986). In the pollen tube of *Rhododendron laetum* grown for 24 hours on an artificial medium, the generative cell envelopes the vegetative nucleus in three long cytoplasmic "tentacles." One of these is the long umbilicus that formerly connected the generative cell to the pollen wall, as described in Chapter 4. Kaul et al. (1987) were, unfortunately, not able to provide a reconstruction after the generative cell divides to form sperm cells. In mature amaryllis pollen (*Hippeastrum vitatum*, monocot family Amaryllidaceae) the lobed vegetative cell nucleus lies close to, but is not connected to, the elongate generative cell.

After pollen germination the vegetative nucleus and generative cell become physically connected and later, after the generative cell divides, the two sperm cells also remain intertwined. In this species the male germ unit forms only after its components are already in the pollen tube. Information about male germ units was included in the general review of pollen biology by Knox and Singh (1987). They summarized information from 18 species in which units have been reported or inferred from micrographs; they also mentioned that other workers have reported that a male germ unit does not occur in some species. In a later review, Mogensen (1992) speculated that isomorphic as well as dimorphic sperm cell pairs could differ from each other in biochemical or organellar characteristics. Alternatively, he speculated that the brief contacts that appear to link sperm cells in some species might be merely an adhesion phenomenon to ensure simultaneous delivery of both sperm cells when the pollen tube bursts after entering a synergid cell.

The miniscule sample of angiosperm species that has been reported to have a male germ unit includes bi-celled and tri-celled species of both dicots and monocots, which suggests that male germ units are widespread but uncommon, similar to the conclusion by Saito et al. (2002) mentioned earlier concerning the taxonomic distribution of dimorphic sperm cells. Three morphological variants of sperm cells are known: identical cells (most common), dimorphic cells (uncommon), and male germ units (uncommon). If there were only identical sperm pairs, the fertilization story would be simpler. The existence of the latter two variants implies that the egg and central cell in some species each requires a "custom fitted" sperm, and raises further questions as to what each sperm provides that is unique and how those characteristics could be built into the sperm cells when the generative cell divides. There is also the possibility that morphologically identical sperm cells are also each different but their unique characteristics are impossible to discern with present techniques, as Mogensen (1992) has speculated. Twell (2002) reported on the small body of information concerning genes expressed exclusively in sperm cells; such studies presently do not shed any light on differences between two paired sperm cells, molecular techniques have the potential to answer some of these intriguing questions.

GUIDING AND NURTURING THE POLLEN TUBE

Much has been learned about pollen tube growth from investigations of pollen behavior on artificial media, a necessary technique for the study of certain aspects. But tubes growing under artificial conditions may not exhibit entirely normal structure or behavior. Kandasamy et al. (1988) have shown that pollen tubes of a tobacco relative (*Nicotiana sylvestris*) growing within the carpel develop a remarkably convoluted plasma membrane, so much so that it appears as a complex network of plasmatubules against the wall of the tube. Pollen tubes of this species grown in an artificial medium, in sharp contrast, have a much simpler, smoother cell membrane. They were able to sample the same tubes in both environments by a simple but clever manipulation.

They germinated pollen on the stigma and later amputated the style just in front of the growing tube tips and placed the cut end down in a gel. The tubes grew seemingly normally into the artificial medium; microscopic preparations showed that tubes in the style had a convoluted cell membrane, but tubes from the artificial medium had an almost smooth cell membrane. Kandasamy et al. (1988) concluded from this experiment that the convoluted membrane greatly extends the surface area of a pollen tube and increases nutrient absorption, and that the convolutions develop only in the natural environment of the carpel.

Their finding suggests that observations on pollen tubes in an artificial medium might not always apply to pollen tubes growing in the carpel. The complex and sometimes contradictory findings regarding pollen tube tip growth, which may partly be the result of these two different research approaches, have received detailed and critical reviews by Steer and Steer (1989) and Mascarenhas (1993).

Pollen tubes in transmitting tissue might grow through some form of viscous substance secreted on the surface of stigmatic epidermal cells and internal epidermal cells of the stylar canal or ovary; in other species pollen tubes burrow instead between cuticle and cell wall or even within the specialized cell walls of transmitting tissue of solid styles. A pollen tube does not penetrate a cell until it enters the embryo sac. A long-held hypothesis is that the pollen tube in the carpel follows a chemical gradient.

The results of numerous experiments on artificial media to test this hypothesis have either been contradictory or open to serious question, as reviewed, for example, by Mascarenhas (1975) and Heslop-Harrison (1987). It might seem that chemotropic attraction is easily tested for by just germinating pollen on the surface of an agar gel and seeing whether the tubes will grow toward some substance or carpel part placed a certain distance away. But there are many possibilities for error in such experiments; for example, tubes might grow toward a cut or chopped-up carpel, but the attraction could be some unrelated stimulatory substance released by tissue damage.

An interesting experiment to test for chemotropism was conducted by Hepher and Boulter (1987) using the self-compatible garden flower annual, painted tongue (*Salpiglossis sinuata*, dicot family Solanaceae). They germinated pollen both on stigmas of intact flowers and on stigmas of isolated styles cut off at the base, and found that the tubes grew downward in both, demonstrating the lack of a long-range attractant from the ovary. Hepher and Boulter then cut stigma/styles and placed pollen grains on the cut basal end of the style, where some germinated and grew back up through the style to the stigma. Fewer tubes grew in this opposite direction, but the fact that any of them did so also demonstrated that there appears to be no long-range downward chemotropic gradient. A similar manipulation by Iwanimi (1959—cited by Heslop- Harrison,

1987) involved removing a section from the middle of a lily style and reinserting it upside down; pollen tubes from the stigma grew normally through this reversed section and into the ovary, a result that also seems to negate the chemotropic gradient hypothesis.

If no long-range chemotropic gradient exists, then perhaps there are local cues instead. The most obvious one is secretion of a chemical attractant from the micropyle or obturator, which would cause a pollen tube that had grown into the ovary and was near the placenta to turn rather sharply and enter the ovule. Mascarenhas (1975) suggested that such local "nudges" might be all that is needed to cause the sensitive pollen tube apex to respond and change its direction of growth.

Another possibility for guidance of the pollen tube is a mechanical track of some kind built into the structure of the carpel. Transmitting tissue that physically (and probably nutritionally) guides tubes. Many species reportedly lack a well-defined secretory track in the ovary, but in many others pollen tubes appear to grow on, or through, a narrow cellular strip or strand that either confines tubes physically or guides them by providing nutrients or other growth substances, or both. This has been well described, for example, in the tepary bean (Lord and Kohorn, 1986).

Observations by Heslop-Harrison et al. (1985) established the following possibilities for the growth of maize pollen tubes:

- Stigmatic trichomes are oriented such that they automatically point most newly emerging pollen tubes towards the ovary; tubes that enter at another angle by accident grow randomly.
- Tubes grow between cortical cells of the stigma until they encounter the transmitting tissue.
- Tubes that are pointed toward the ovary when they enter the transmitting tissue simply continue growing in that direction, although some might become diverted by accident into the stigma cortex and grow there randomly.

- Tubes that reach the ovary grow over epidermal cells elongated in the direction of the single ovule.
- Tubes that become diverted accidentally in the vicinity of the micropyle grow randomly and might even start back toward the stigma.
- The single successful tube turns toward the micropyle and enters it.

A naturally occurring mutant provided evidence for mechanical rather than chemical guidance of pollen tubes. A mutant form of pearl millet (*Pennisetum typhoides*) lacking stigma trichomes allowed Heslop-Harrison and Reger (1988) to test pollen tube behavior in the absence of these cells, which normally seem to orient their direction of growth. Emerging pollen tubes entered the pearl millet stigma at the base of the non-papillose epidermal cells of the mutant, but lacking the physical orientation provided by the trichomes, they grew at various angles and directions within the stigma, with some tubes reaching the transmitting tissue only by accident. The investigators concluded that the stigma trichomes of pearl millet, maize, and probably many other grasses provide at least an initial mechanical orientation and guidance for pollen tubes.

Wiesenseel et al. (1975) demonstrated that lily pollen tubes drive a steady electrical current of a few hundred picoamperes. This amounts to a flow of positive ions, a current that enters at the tip of the tube and exits in the grain. Because potassium ions can enter anywhere in the tube, but calcium ions enter only at the tip, the electrical field appears to be driven by the flow of calcium ions. This can perhaps be related to an earlier popular idea that a calcium gradient is important in guiding the pollen tube. Investigators who work with a particular species usually do so in the hope that what they discover will apply generally to all species. The structural, chemical, and physiological variation that occurs among pollen and carpels, of which some examples have been given in this and previous chapters, might make it impossible (as yet, at least) to reach a grand all-encompassing generalization about pollen tube and carpel relations.

CALLOSE PLUGS

The living vegetative cell occupies a certain volume of the tip of the lengthening pollen tube, although behind it is a mostly empty shaft, as in the descending elevator car analogy mentioned earlier. Transverse barriers typically deposited at intervals in the tube consist mostly of callose, the all-purpose sealer that plants produce in so many different places. Examples of these callose "plugs" can be seen at low magnification in sweet cherry (*Prunus avium*) pollen tubes growing in the style. These plugs were first observed in 1878, and by 1890 they had been identified as callose. Brink (1924) experimented on pollen tubes, demonstrating that severing a tube behind the newest plug did not inhibit tube growth, but cutting in front of the newest plug ruptured the vegetative cell and killed it.

This verified that the older part of the tube, subdivided by periodic callose deposits, is merely a series of remnants. Brink (1924) said that the first plug is usually deposited at a distance of 400–700 ¼m from the grain, and that new plugs are formed no closer than 200 ¼m behind the tip.

According to Vasil (1987) the first plug always appears at a fixed distance behind the tip, depending on the species, and it never forms until the tube is at least 1 mm long. Callose plugs are not always deposited at fixed intervals along the tube, however, as is evident from and many similar published illustrations. Bassiri et al. (1987) illustrated, and commented on, the "irregularly spaced and intermittent" callose plugs formed during in vitro and in vivo selfand cross-pollinations of three chickpea species (*Cicer*, dicot family Fabaceae).

In an investigation that utilized petunia pollen germinated both in vitro and in vivo, Mulcahy and Mulcahy (1982) found that tubes in the petunia carpel grew slowly for the first 7–11 hours and formed no callose plugs. They interpreted this to be an "autotrophic phase," which utilized stored pollen food. The growth rate then quickened to about three times faster, and callose plugs formed periodically in the tubes. This they interpreted as the "heterotrophic phase," when nutrients were being absorbed from the carpel. Pollen that germinated in vitro grew only for about 11 hours, and only attained about 10% of the length of in vivo pollen tubes grown for 11 hours.

In vitro tubes formed callose plugs only after they stopped growing. These results suggest that callose plugs are formed only after the vegetative cell has become dependent on maternal host tissue for nutrition. Further support for this two-phase concept of pollen tube growth and its association with callose plugs came from a later amplification of this study (Mulcahy and Mulcahy, 1983), in which they found almost the same growth results for tomato pollen tubes. Petunia and tomato have bi-celled pollen. Campion (*Silene dioica*, dicot family Caryophyllaceae) has tri-celled pollen. Mulcahy and Mulcahy (1983) found that pollen tubes of campion exhibited a growth curve corresponding only to the heterotrophic mode, and that callose plugs were laid down much earlier than in petunia and tomato. They felt that campion is heterotrophic from the start, which explains both its faster growth and the greater difficulty of germinating tri-celled pollen in simple artificial media.

Why are callose plugs deposited? Brink (1924) ventured three reasons:

- The plugs maintain the integrity of the vegetative pollen cell.
- The plugs prevent absorption of nutrients from already traversed and therefore "exhausted" parts of the style.
- If the tube stops growing and goes dormant, as it does over the winter in certain temperate zone trees, the plug helps to turn the vegetative cell into a cyst.

Vasil (1987) opined that callose plugs serve to restrict the living protoplast to the tip region. He also suggested that the plug closest to the tip helps maintain turgor pressure by preventing the osmotically active vegetative cell from expanding in a rearward direction. Pressure directed toward the tip, however, where the newly forming tube wall is thinnest, causes the wall to be pushed forward and stretched, thereby providing a motive force that moves the vegetative cell. As the callose plug is left further behind, this osmotic force weakens, and a new plug must form closer to the tip to reinstate the pressure. The callose plugs can be viewed, perhaps in a simplified way, as a series of immovable barriers against which the vegetative cell braces itself in order to move itself forward. The observations on petunia by Mulcahy and Mulcahy (1982) mentioned earlier support the idea that faster growing tubes need more frequent plugs. It must be stated, however, that in many pollen tubes callose plugs have been reported to be infrequent or absent, which weakens this otherwise attractive hypothesis.

SWELLING AND BRANCHING OF POLLEN TUBES

The vegetative pollen tube cell moves through the carpel by progressive growth of its wall, probably helped by occasional or regular deposition of callose plugs to help push the cell forward. This hypothesis needs to be related in some way to the two common phenomena of swelling and branching of pollen tubes. Swelling of the tip is frequently associated with arrested growth of the pollen tube as part of an incompatible reaction. Incompatible tubes in general have larger amounts of callose within them than do compatible tubes, according to Vasil (1987). If the more mature parts of the tube are tougher because of more callose in the wall, and if incompatibility renders the tube incapable of softening the transmitting tissue ahead of it sufficiently to allow continued penetration, a last growth response just before the vegetative cell dies could be a lateral expansion of the new, thin, and therefore most extensible part of the wall just behind the tip.

Branched pollen tubes were reported by many earlier investigators. Steffen (1963b) regarded these as "abnormal phenomena" and, while not denying that they occur, remained skeptical of explanations as to their possible function. Maheshwari (1950) reported several examples of branched tubes, as well as persistence of the main tube, and speculated that the purpose of such branches is to absorb additional nutrients for embryo development. There is, however, no evidence for any function of such branches. Striking and irrefutable evidence that branched pollen tubes occur under natural conditions was shown by Wilms (1974) for spinach. Although some branched tubes were seen on or in the stigma, he reported that branching occurred most frequently in the ovary, ramifying around the micropyle and between the integuments, but always on the surface of the ovule. The tips of some tubes appeared amoeboid, with an irregular terminal swelling and several extrusions. Scanning electron micrographs of the inner surface of the radish ovary by Hill and Lord (1987) show irregular, knobby, and branched tubes zigzagging their way along the grooves between epidermal cells.

None of the published accounts that mention or show branched pollen tubes include any information about whether the branches contain protoplasm or callose plugs. If the vegetative cell is wending its way toward the ovule as a single cell with nucleus and two sperm cells, how can it subdivide itself in order to move into two or more branches? The long side branches shown in spinach, and the other examples reported from other plants by earlier workers, must form by the activity of an organized protoplast, but how can the vegetative cell fragment itself in some amitotic fashion into several living subdivisions? No answers have been provided as yet to explain branched pollen tubes.

POLLEN TUBE COMPETITION AND CARPEL "FILTERS"

There is considerable evidence that among any group of compatible pollen grains that germinate and begin to grow normally, the various pollen tubes differ in their growth rates because of different genetic characteristics. If the number of compatible pollen tubes growing in a style exceeds the number of ovules available to be fertilized, only the fastest-growing tubes will be able to pass on their genes to the seeds. In addition to competition by different growth rates, at least some gynoecia have structural bottlenecks or even more subtle mechanisms that filter out all but the appropriate number of tubes. In this section, some examples of these mechanisms are described.

Mulcahy (1986) and Mulcahy and Mulcahy (1987) reviewed the evidence bearing on whether pollen tube competition exists and, if so, what is its significance. Without going into details, much of which involves genetic aspects beyond the scope of this book, they suggested that pollen tube selection operates in species that regularly receive more pollen at one time on the stigma than there are ovules. The Mulcahys give a number of specific examples of how this selection works, based on experimental treatments favoring one or another of the competing pollen grains and pollen tubes. A straightforward example from Mulcahy and Mulcahy (1987) illustrates one possibility.

The carnation (*Dianthus chinensis*, dicot family Caryophyllaceae) has a pair of very long slender stigmas, each of which resembles a style. Pollen may land and germinate anywhere between the tip and the base of this elongate stigma, which means that a pollen tube starting at the tip must grow at least twice as far as a tube starting from the base. In one experiment, pollen was placed only at the stigma tip in some plants (long distance: more competition) and only at the stigma base in others (short distance: less competition and greater possibility for chance fertilizations). Seeds resulting from both of these controlled pollinations were similar in size and appearance, but seeds from stigma tip pollination germinated more rapidly and grew faster than did seeds from stigma base pollination. The Mulcahys interpreted this difference in seed germination as the result of pollen competition, which allowed the fastestgrowing pollen tubes to incorporate their genetic traits in the next sporophyte generation.

The competition among growing pollen tubes is intensified in some species by one or more filters, mechanical or otherwise, that reduce the number of tubes. The best examples are from grasses. Heslop- Harrison et al. (1985) described the structural bottlenecks that exist in maize (the pollen tube pathway in general in maize was described in the earlier section on pollen tube guidance).

At the base of the maize stigma the transmitting tissue passes through an abscission zone just above the stigma's junction with the ovary. Just 6 hours after pollination, cells in this area begin to lose turgidity, and by 24 hours the entire zone is flaccid and the internal tissues have become disrupted, which blocks the passage of any further tubes. In one demonstration more than 30 pollen tubes approached the abscission zone but only 5–10 managed to grow beyond it. The transmitting tissue tract narrows when it approaches the inner epidermis of the ovary, creating a constriction that eliminates any additional laggard tubes. Only a few pollen tubes penetrate the cuticle of the inner ovary epidermis, from which point they grow somewhat erratically toward the single ovule. The micropyle appears to secrete a polysaccharide-rich attractant, and only one tube enters the ovule.

Another example of a constriction in the pollen tube pathway occurs in the upper portion of the style of tepary bean, as described by Lord and Kohorn (1986). The spacious canal that extends upward into the style from the ovary ends abruptly, and solid transmitting tissue occurs between it and the stigma surface. Many tubes grow through the solid stigma and enter the stylar canal, but they can only grow along the ventral side, which is lined by inner epidermal cells that secrete substances to form a narrow track. This seems to act as a bottleneck type of filter that favors the fastest-growing pollen tubes. In sweet cherry, on the other hand, tubes encounter a constriction at the base of the style, where the broad column of transmitting tissue tapers in such a way that the first few tubes can grow through but all later ones are arrested just above it, where their tips swell and burst (Raff et al. 1981).

There is a much more subtle pollen tube selection process in the avocado (*Persea americana*, dicot family Lauraceae). This species has only one ovule per ovary, which normally contains only one embryo sac, but some ovules develop two embryo sacs. Sedgley (1976) examined 74 avocado carpels after pollination (a very large sample in this kind of work!) and found the following progressive pollen tube mean values per carpel: 66 pollen grains germinated on the stigma but halfway down the style only 28 tubes continued to grow, and only three or four tubes were noted at the base of the style. An ovary with one embryo sac in the ovule received only one pollen tube, but in five ovaries the ovule had two embryo sacs, and each of the five received two pollen tubes. Essentially the same results were obtained when additional carpels were examined from two varieties of avocado (Sedgley, 1979). Sedgley concluded that the avocado embryo sac (or sacs) exerts an unknown influence, which either selects one (or two) tube(s) or else inhibits all but one (or two). Further work is called for, but the large sample size and careful methods used here inspire confidence in Sedgley's conclusion that an unknown selective factor gradually eliminates pollen tubes between stigma and ovule.

Stösser (1986) tested for a possible attractive influence of the ovary on pollen tube growth by germinating pollen both on intact carpels and on isolated stigma/styles removed from apple, cherry, pear, and plum flowers (all Rosaceae). He found no differences in rates of pollen tube growth or appearance of the tubes in control or experimental tissues. In addition, in some varieties of tomato he found that extra styles were produced in which the transmitting tissue did not extend to the ovary. Tubes in these styles grew normally to the style base, but stopped growing there. He also noted that in control flowers, several tubes often entered an ovary and grew in a zigzag and seemingly random manner, which indicated little or no attraction by the ovule. His conclusion from these experiments was that neither ovary nor ovule exerted any influence on the pollen tubes in the species investigated.

RATE AND DURATION OF POLLEN TUBE GROWTH

The previous section included a discussion about pollen tube competition, with examples showing that tubes differ in their growth rates, even within one species. But there was no mention of how fast pollen tubes actually grow or how long it takes for tubes to grow from stigma to ovule. Both of these aspects have been studied many times. A few examples are presented here.

It was mentioned earlier in this chapter that Mulcahy and Mulcahy (1982, 1983) showed that bi-celled pollen grew at a slow rate during their early, autotrophic phase, and at a faster rate during their heterotrophic phase, as in petunia. Tri-celled pollen grew at close to the heterotrophic rate from the beginning, which they suggested was inherent in such pollen. Hawkins and Evans (1973) reported that pollen tubes in two bean species (*Phaseolus*) grew faster in transmitting tissue if they were close to the vascular tissue, slower if they were on the periphery. This is another possible variable that in some species could affect the rate of pollen tube growth. Most other investigators have not made fine distinctions within the species they studied.

The average rates of pollen tube growth reported from among several cultivated species ranges from about 0.3 mm per hour to somewhat more than 6.0 mm per hour. Some examples (in mm/hr) are: apple (0.35), petunia (O.3-autrophic, 0.8-heterotrophic), lily (0.9), bird-of-paradise (1.8), bean (2.0), iris (4.0), rye (5.4), and corn (6.25). Maheshwari (1950) concluded that pollen tubes of most species take between 1 and 48 hours from germination to ovule entry, without considering the different carpel distances.

Cereal grasses in general take a short time, from 1–3 hours being common, although rice requires about 12 hours; so does the coffee tree. Lettuce needs only 3–5 hours, but spinach pollen tubes take 7–14 hours. Most herbaceous legumes require about 24 hours, as do some poplars. Many woody plants take much longer times—e.g., pecan at 4–7 days and apple at 8–12 days. Among extreme examples, the time required is measured in weeks or months, especially in certain temperate zone trees where pollination occurs in one season, the tubes go dormant for the winter somewhere between stigma and ovule, and then resume growth in the next year. Experimental studies favor species with fast-growing pollen tubes; thus little is known about what happens when tubes grow very slowly or go dormant en route.

POLLEN TUBE GROWTH IN OVARY AND OVULE

Pollen tubes have been reported by some investigators to grow randomly in the ovary—e.g., in rosaceous fruit tree species (Stösser, 1986) and in the ovary of spinach. The few grasses studied seem to have some mechanical guidance in the ovary, but there is no track consisting of secretory cells (Heslop-Harrison et al., 1985, and see earlier section on guiding the pollen tube). Species in some other families have been shown to have transmitting tracks in the ovary that guide pollen tubes directly to the micropyle, or at least to an obturator that surrounds the micropyle. Such a direct pollen track has been described in the cultivated fig (*Ficus carica*, dicot family Moraceae) by Beck and Lord (1988). Fig pollen tubes grow through solid transmitting tissue all the way to an obturator located on the funiculus of the single ovule. Tubes growing into the ovary encounter secretion from the obturator, which lifts the cuticle. Tubes grow through this secretion and into the micropyle.

Wild radish (*Raphanus raphanistrum*) pollen tubes grow to various levels in the ovary in a strip of transmitting tissue within the septum that divides the two carpel ovaries. Tubes exit at various places through the epidermis of the septum, grow over the septum surface and across an obturator, and then enter the micropyle of an ovule.

The tepary bean (*Phaseolus acutifolius*) has a typical elongate leguminous ovary with two narrow placental ridges to which the ovules are alternately attached. A pollen tube growing on one of these cuticle-free ridges encounters a micropyle that almost touches it, which provides convenient entry without an obturator. In the ovary of the ornamental monocot, bird-of-paradise (*Strelitzia reginae*), the placenta is a ribbon of secretory tissue, and pollen tubes are restricted to its surface (Kronstedt et al., 1986). Anatropous ovules with obturators occur in two rows along this placental strip and their micropyles almost touch it, which offers easy entry for pollen tubes.

The pollen tube does not always enter an ovule through the kind of wide-open micropyle typically shown in textbooks. In Chapter 6 the micropyle was shown to often occur as a crack or fissure that gives the impression of requiring some force to open it wide enough to accommodate a pollen tube. There is a contradictory body of information on chemotropic attraction by the ovule, especially by the micropyle; the evidence was reviewed by Steffen (1963b), who doubted that micropylar attraction occurs. A recent study indicates that an active secretion emanates from a synergid cell to attract a pollen tube to the ovule of the garden ornamental, *Torenia* (dicot, Scrophulariaceae) (Lord and Russell, 2002).

Direct evidence for the occurrence of a fluid in the micropyle that could attract pollen tubes was provided by Chao (1971, 1977, 1979) from his work on two forage grasses, *Paspalum longiflorum* and *P. orbiculare*. Chao showed that a conspicuous amount of water-soluble mucilaginous substance fills the spaces between integuments, between the outer integument and the ovary wall, and also between columns of nucellar cells below the micropyle.

The secretion is composed mostly of an amorphous polysaccharide permeated by minute fibrils, and it seems to be secreted both by nucellar cells and by some cells of the integuments. Chao concluded that this substance attracts pollen tubes and also provides nutrients for tubes growing toward the embryo sac. It is also possible that the synergids could contribute to this secretion, as just mentioned for *Torenia*. The examples in this section, like those from the embryological literature in general, show that pollen tracks are lacking in some ovaries whereas others provide rather precise pathways to the micropyle. The evidence indicates that an ovary with only one or two ovules usually lacks a secretory track, although *Paspalum* appears to be an exception.

POLLEN TUBE DISCHARGE AND DOUBLE FERTILIZATION

The three diagrams of Jensen (1972) combine information from many sources to provide a generalized scheme for pollen tube entry into the embryo sac. The tube penetrates the filiform apparatus of one synergid (which may have already begun to degenerate), and in the synergid cell the intrusive pollen tube tip bursts and expels the sperm cells; the sperm cells then contact the egg cell and central cell membrane, respectively, shed their scanty protoplasts, and only the nuclei enter their respective cells. What propels the sperm cells to these two different cells, and how they subsequently make their way within these cells to the target nuclei, remains to be resolved.

As an actual example, the pollen tube of Iceland poppy (*Papaver nudicaule,* dicot family Papaveraceae) can be seen after it has grown over the ovule and is entering the micropyle of the anatropous ovule; a closer look within the ovule shows the pollen tube after it has squeezed between cells of the nucellus and penetrated a degenerating synergid cell via the spongy, channel-filled filiform apparatus. Entry of the tube either causes the synergid to start degenerating, or the synergid has already begun to decline before the tube reached it. There are reports from other species of syngergid degeneration happening either way. Some investigators have suggested that a substance might exude from a precociously dying synergid and attract the pollen tube.

It may be that the synergid always starts degenerating early, not necessarily to secrete something, although there is evidence for that (see earlier section on attraction of pollen tube to ovule), but because its loss of turgidity might make it a soft and easily penetrated spot for the pollen tube. The persistent synergid in Figure 8.16B is adjacent to the degenerated one, and its nucleus is just above the darkly channeled filiform apparatus. Whatever the condition of the synergid, events following pollen tube entry into it have not been easy to interpret, because the protoplasm from the discharging tube mixes with that of the degenerating synergid and obscures the sperm cells and nuclei. Both figures show that the degenerating cell contents obscure distinguishable features within the synergid.

Speculation as to why a pollen tube seems to always enter the embryo sac by penetrating a synergid must consider the peculiar structure of the filiform apparatus, and perhaps also the loss of synergid turgidity if the cell has already begun degenerating; seepage of degenerating synergid cytoplasm through the filiform apparatus wall channels could be a source of pollen tube attraction. The tip of the pollen tube ruptures ("bursts open" according to some investigators) soon after it enters the synergid and expels the sperm cells. This is the first cell that the pollen tube actually penetrates, but if the synergid has already died or is degenerating, perhaps it is not at all the same as puncturing a turgid living cell. Some have speculated that the osmotic conditions within the synergid cytoplasm are different enough from that of the pollen tube that this causes the tip to rupture.

From the moment that sperm cells are expelled into a synergid to the moment when they contact the egg cell and central cell nuclei, respectively, details are unclear. It must be appreciated that these are short-lived events, the entities involved are vanishingly tiny, and the cytoplasm is murky. The movements and merger of nuclei during double fertilization have mostly been shown as interpretive drawings, which does not mean that they are incorrect, but rather that the process is very difficult to follow and interpret, and that microscopic preparations of the embryo sac during these stages are rarely of photogenic quality. Among the few accounts of double fertilization that have been presented photographically in some detail, that of Luxová (1967) on barley is worth reproducing here. Luxová collected a sample of 15–20 carpels every 10 minutes following pollination in the field, and repeated her sampling during a second year. Each carpel was sectioned for light microscopy, and individual sections that included the best information were photographed.

The first set of photographs shows the pollen tube in the embryo sac only 20–30 minutes after pollination (A) and its discharge and expulsion of sperms (B,C), of which one sperm is seen in contact with the egg cell membrane and the other sperm is shown partway toward the polar nuclei. The sperm in transit to the polar nuclei is also shown in (E). The sperm in contact with the egg moves into the cytoplasm and contacts the nucleus (D) only 60–70 minutes after pollination, before the second sperm has even reached the polar nuclei.

The second series of photos documents the gradual merger of sperm nucleus and egg nucleus. Note that after merger the nucleus (now the zygote nucleus) appears to be in a prophase-like state (I). The third set of pictures shows the second sperm progressively merging with the two polar nuclei. In barley, the two central cell nuclei typically do not fuse with each other before they merge with the sperm nucleus. Figure 8.21I jumps ahead to show the end of the first division of the primary endosperm cell nucleus, which thereby initiates the coenocytic phase.

The excellent work of Luxová and others using light microscopy has not been able to determine the fate of the cytoplasm of the sperm cells, which is scanty and therefore difficult to identify. The question of whether sperm cell cytoplasm, and particularly any mitochondria or plastids, is carried into the egg cell has long been debated. Sperm cell plastids of some plants have been reported to enter the egg cell and to persist, mingled with maternal plastids, up to the globe stage of the embryo, but there is no report of such biparental inheritance of plastids extending beyond the globe stage.

It shows that the cytoplasm of one of the sperm cells is left behind in the degenerating synergid while its nucleus continues on into the egg cell. Mogensen suggested alternatively that perhaps the "cytoplasmic body" represents the fused cytoplasm of both sperm cells, since he could find no evidence of the fate of the cytoplasm of the second sperm nucleus on its way to the central cell nuclei.

The male germ unit and dimorphic sperms were described earlier in this chapter, and they must now be reconsidered with respect to fertilization. Until recently it was assumed that the two sperm cells were identical, each having an equal chance of merging with egg nucleus or central cell nuclei. A review of fertilization by Knox and Singh (1987) considered the question of whether there is a true male sperm, one that is predestined to go to the egg. They analyzed the work of Russell, whose study of *Plumbago* discussed earlier revealed a conspicuous difference in plastid and mitochondrial populations between the two sperm cells.

They concluded that Russell was correct to conclude that the plastid-rich sperm cell is destined to fuse with the egg; however, they added that the evidence does not show conclusively that the mitochondria- rich sperm will always go to the central cell. This remains a crucial unanswered question. How sperms move within the embryo sac is speculative. Amoeboid and undulatory motions have been proposed, as well as undefined fields of attraction. Cytoplasmic strands that are known to cross the central cell vacuole could possibly carry one sperm to the polar nuclei. Microtubules or microfilaments could possibly guide the sperm nucleus inside of the egg cell, and F-actin strands have been implicated in recent molecular studies (Lord and Russell, 2002).

Isolation of sunflower embryo sacs by enzymes or micromanipulation after pollen tube entry allowed Zhou (1987) to view fertilization directly under the microscope and to follow events up to proembryo and multinucleate primary endosperm cell. Sperm cells were elongate in the pollen tube but became spherical, oval, or spiral in the embryo sac.

The techniques for isolating sperm cells and embryo sacs are being developed, and perhaps when fertilization in isolation is possible, these and other technically very difficult questions may be answered. The observations by Zhou (1987) of fertilization in sunflower ovules isolated after entry of the pollen tube show that such studies are possible.

POLYSPERMY

Sperm cell movement in the embryo sac is difficult to follow when only one pollen tube enters, and more so when two or more pollen tubes enter the same ovule. Figure 8.18 provides direct evidence of this phenomenon, which is one kind of polyspermy ("many sperms"). Also included under polyspermy is the unusual phenomenon of more than one mitosis of a single generative cell, which is another way to produce extra sperms for the embryo sac.

A review of polyspermy by Vigfusson (1970) gathered 55 reports from the literature (35 dicots and 20 monocot species), many of them common cultivated plants such as cotton and maize. All reports were from outcrossing species; polyspermy seems not to occur in self-pollinating species.

Fertilization of the egg by sperm from one pollen tube and polar nuclei by a sperm from a second pollen tube has been demonstrated cytologically in maize, and termed "hetero-fertilization" by Sprague (1932). Steffen (1963b) regarded polyspermy as an abnormal phenomenon, but Vigfusson (1970) concluded that it is widespread among crossfertilizing angiosperms, and implied that it is too common to be dismissed as an abnormality.

He gave no reason, however, why such occurrences could be considered to have any advantage. Normal or abnormal, there seems to be no explanation available as to why such a phenomenon occurs. One could speculate that if two pollen tubes arrive by accident at the same micropyle, and if it is open enough to accommodate them, then one could imagine both entering the embryo sac and discharging sperms, whereupon confusion could occur from having four sperm cells and nuclei.

9

Endosperm

Double fertilization initiates both embryo and endosperm. Although inseparable in the developing seed, they are always described and discussed separately because each is a major topic. But because endosperms and embryos must be shown together, there is some inevitable cross-referencing. The custom is to consider endosperm first, probably because in almost all species it begins to develop first, and often proliferates considerably before the zygote divides. This chapter also mentions perisperm, a lesser-known nutritive tissue that proliferates from the nucellus in some dicots and monocots and supplements or replaces the endosperm.

GENERALIZATIONS AND HISTORICAL INTERPRETATIONS

Endosperm merely means "within the seed," a nonspecific term coined in the middle of the 19th century when little was known about its initiation and development. The discovery of double fertilization stimulated considerable speculation about how to interpret this puzzling pair of events. Some early investigators made the plausible interpretion that endosperm represents a potential second embryo that evolved into a nutritive entity, but Trelease (1916) concluded that endosperm is neither gametophyte nor sporophyte, but something coequal to both; he called it the "xeniophyte" (Greek for "hospitable plant").

Friedman (1998) reviewed the history of studies on endosperm evolution and, combined with original work on reproduction in gymnosperms, became convinced that endosperm was indeed originally a second embryo, which in angiosperms became a nurturing tissue for the primary embryo. This interpretation, however, like all others, remains speculative. Because endosperm is initiated by an event that seems equivalent to fertilization, it could be interpreted as part of the new sporophyte, but no text seems to espouse that idea.

Because of its unusual development and ultimate fate one can appreciate the difficulty of classifying it, and it is easy to see why Trelease suggested that it be called something entirely different.

Early investigators correctly described fusion of the second sperm with the so-called polar nuclei, but they were uncertain of where the merged nuclear trio resided in the embryo sac after the antipodal cells and egg apparatus cells were subtracted. The first embryology text (Coulter and Chamberlain, 1903) described the polar nuclei as residing "...in the general central region of the sac, forming the primary endosperm nucleus."

This rather vague description was continued in the standard works of Maheshwari (1950, 1963), although Brink and Cooper (1947) had clearly stated, "The central cell of the female gametophyte is fertilized by the other sperm and develops into the endosperm." They also used the term "primary endosperm cell" for the fertilized central cell. Jensen (1972) said that this is possibly "...the only case in which the nuclei of a cell have a name while the cell itself has gone essentially nameless." Although the central cell has been recognized subsequently by Willemse and van Went (1984) and Vijayraghavan and Prabhakar (1984), and others, some still refer just to the nuclei and forget the central cell.

The nomenclature introduced here for types of endosperm recognizes that the fertilized central cell, the primary endosperm cell (PEC), is the originating entity. Emphasis on the nucleus has led to misleading names and concepts about some pathways of endosperm development. New terms are proposed in this chapter to clarify the names for three of the four types of endosperm.

The primary function of all types of endosperm is to provide nutrients to the developing embryo, and often to the seedling as well. Four other endosperm functions are also known, which will be taken up near the end of this chapter. Endosperm nourishes many animals, and it is almost certainly the main food source for humans, either processed directly from plants, such as cereal grasses, or indirectly from endosperm transformed and transferred to the embryo cotyledons of plants, such as legumes. Kowles and Phillips (1988) reported that about 380 billion pounds of endosperm are produced every year in the United States alone. Such commercial importance stimulates research, and there is now an enormous amount of published information on endosperm and related aspects of cotyledon storage in many food plants. These aspects can be touched upon only briefly here and in the next chapter.

CYTOLOGY OF ENDOSPERM

In most angiosperms, two nuclei and one sperm nucleus merge in the central cell to form the triploid nucleus of the PEC, but there are exceptions; the central cell of a small number of species has only one nucleus, and some have three or more. Fertilization raises the initial DNA level of a typical PEC nucleus to triploid level, which probably enhances in some way its ability to attract nutrients. Much higher DNA levels are common in later-formed endosperm cells as a result of atypical mitoses or unusual amitotic variations (Vijayraghavan and Prabhakar, 1984) that produce polyploidy and endopolyploidy. These elevated DNA levels range from modest increases to more than a thousand times original amounts in some species.

Such proliferation probably enhances the ability of endosperm to compete for nutrients, but it is also a gross genetic disturbance that allows endosperm to exist only as an ephemeral storage tissue (except for the modified peripheral one to a few endosperm cell layers of many species). Similar abnormal nuclear behavior occurs in cells of the proembryo suspensor, another ephemeral structure commonly developing elevated DNA levels that has evolved in several, sometimes seemingly bizarre, ways. Cells of the short-lived anther tapetum also commonly have raised DNA levels, although not to the very high levels known from suspensors and endosperm. All three of these entities are involved with nutrition, among other functions.

Endosperm cells that remain undigested at seed maturity may be either alive or dead, depending on the species or family. Grasses, for example, have dead mature endosperm cells, whereas they remain alive in many other groups of plants. The significance of these differences in vitality has yet to be explained. The peripheral endosperm cells called aleurone cells are, however, reported to be living in all studied species (DeMason, 1997).

INTRODUCTION TO ENDOSPERM TYPES

All endosperm is cellular, regardless of how it develops, because all protoplasm must be contained within at least an intact cell membrane, even if a wall is absent. After merger with a sperm cell nucleus, the PEC grows into some form of mature endosperm tissue by one of four possible developmental pathways, referred to as "types." The PEC almost always begins growing before the fertilized egg divides; in some species it is only a slight lead, but in most there is noticeable to considerable endosperm first, and it might grow for as long as several weeks before the zygote divides.

In several families the PEC simply undergoes mitosis and cytokinesis to produce a multicellular tissue in a straightforward way. This type has been called "ab initio cellular endosperm" and "cellular endosperm." Because all endosperm is cellular, as emphasized earlier, this type will here be called "multicellular endosperm" to indicate that no coenocytic phase occurs.

The most common pathway of endosperm development is not so simple, however, because the PEC first enlarges greatly and becomes a multinucleate coenocyte, which only later subdivides itself partially or completely into uninucleate cells by internal wall formation. Additional endosperm cells usually proliferate later by either ordinary or variant forms of mitosis followed by cytokinesis.

This type is usually called "nuclear endosperm" but that is not the best descriptive term because it gives a noncellular connotation to what is really the multinucleate PEC, and because "nuclear" describes only the first growth phase of such endosperm. This pattern of two-step endosperm development can be more accurately labeled as the "coenocytic/multicellular" type. This is not the first use of the term coenocyte for a multinucleate PEC; Newcomb (1978) used it, for example, to describe the two multinucleate endosperm cells that develop first in an amaryllis relative, the African blood lily. A variant of the coenocytic/multicellular type is the so-called "Helobial endosperm" in which the PEC first divides to form two unequal uninucleate cells.

The largest cell next to the zygote (micropylar endosperm cell) then follows the coenocytic/multicellular pattern; the smaller chalazal endosperm cell typically develops into a coenocyte that never subdivides itself into cells. One could say that two types of endosperm develop as concurrent subtypes within the Helobial type. This name has been retained because it would be cumbersome to devise a more precise descriptor, and because most known examples are from the monocot Order Helobiales.

In a small number of plants of various families, the PEC enlarges and produces numerous nuclei, but remains as a coenocyte and never subdivides itself. This type has also been called "nuclear endosperm," a term that ignores the fact that these nuclei are within a cell. This type will here be called "coenocytic endosperm." It should be mentioned that some or all species in a few families either never form endosperm or it is initiated but degenerates almost immediately. The only family in this category of interest here is the orchid family (Orchidaceae), where many species have no endosperm and their embryos also remain rudimentary (but often develop bizarre suspensors) within the tiny seeds.

The four pathways by which the PEC may develop into endosperm are illustrated diagrammatically in a morphological (not evolutionary) sequence, and each is described in more detail in the following sections. The last section of this chapter speculates about the possible advantages of different endosperm types.

MULTICELLULAR ENDOSPERM

The multicellular endosperm is the most easily understood type. The PEC never becomes a coenocyte but instead divides, as do cells elsewhere, by mitosis followed by cell plate deposition, thereby directly producing two endosperm cells. Each of these cells divides similarly, and this continues until a certain number (hundreds or thousands) of endosperm cells have formed. This straightforward pathway of endosperm development occurs in 79 families, all except two of which are dicots (Davis, 1966), and the clustered distribution of these families has convinced some theorists that it is the original endosperm type from which the other types have evolved (Dahlgren, 1980).

The Solanaceae with its many cultivated species is one economically important dicot family with multicellular endosperm. Several studies have established that this type occurs among such cultivated species as potato, tobacco, and tomato. A good example is *Nico tiana rustica* (Cooper and Brink, 1940), a tobacco species representative of the family. The PEC divides shortly after fertilization and there are two endosperm cells while the zygote remains undivided. When the zygote finally divides, there are already 16–24 large endosperm cells in the expanded embryo sac. Endosperm cells continue to divide as the proembryo enters the globe stage, multicellular endosperm at 5–6 days after fertilization.

Endosperm cells continue to proliferate until much later, but at the same time the growing embryo is also expanding and progressively eliminating individual endosperm cells for nutrition. The reproductive calendar of tomato, also Solanaceae, correlates multicellular endosperm development with other post-fertilization events in an easily grasped chart form.

COENOCYTIC/MULTICELLULAR ENDOSPERM

The coenocytic/multicellular endosperm is by far the most common type, known from 161 families, of which 83% are dicots

(Davis, 1966). The PEC first enlarges and proliferates nuclei, often by synchronous waves of mitosis, thereby becoming a coenocyte. The number of nuclei produced is at least approximately characteristic for a species, ranging from eight up to many thousands. In wheat and barley, for example, over 2,000 PEC nuclei are present before any internal walls form (Bennett et al., 1975), and about 3,000 nuclei have been reported in the PEC coenocyte of apple before it subdivides itself into cells (Wanscher, 1939).

The coenocytic PEC enlarges, and among different species it produces a few to numerous nuclei even before the zygote divides. When the embryo does begin to grow, it eventually impinges on the expanding coenocytic PEC and becomes partially to almost completely surrounded by it. The PEC cell membrane remains intact, as it must to retain its cellular integrity. The PEC can accommodate its increasing volume to the intrusive embryo as well as to the surrounding nucellus because it expands its cell membrane and also remains rather loosely attached to the embryo sac wall. Yeung and Cavey (1988) showed a meshwork of loose fibrils between the PEC and the embryo sac wall in the common bean (*Phaseolus vulgaris*). They were able to dissect out an intact PEC with its surrounding wall from the embryo sac, demonstrating that the coenocytic PEC is a true cell. Yeung and Cavey (1988) also showed conspicuous folds in the PEC wall, an indication that the PEC is expandable.

PEC nuclei are not dispersed haphazardly. The enlarging central vacuole pushes the cytoplasm with its nuclei to the periphery of the PEC, including the peripheral zone adjacent to the growing embryo. Microtubules appear later in the peripheral PEC cytoplasm and arrange the nuclei spatially, which determines the position of future endosperm cell walls (DeMason, 1997). After a certain number of nuclei have proliferated and are regularly arranged, the PEC begins subdividing itself into cells. These early endosperm cells often form first adjacent to the embryo, as for example in *Phaseolus* (Yeung and Cavey, 1988). In other species, the first cells form around the lateral periphery of the PEC, as for example in the

African blood lily (Newcomb, 1978), and cells may even form first at the chalazal end, as in the monocot *Iris sibirica* (Iridaceae) (Olszewska and Gabara, 1966).

It is difficult to visualize precisely how these internal cells form within the PEC coenocyte. A superficial description is that a series of cell walls form progressively inward from the PEC periphery and, after nuclear divisions occur, cross walls form, completing each box-like cell and enclosing one nucleus. Van Lammeren et al. (1996) provided a detailed but readable description for rapeseed (*Brassica napus*, Brassicaceae), and DeMason (1997) reviewed several published studies of wall formation and pointed out disagreements among the descriptions. The inconsistencies in published accounts attest either to the difficulty of accurately describing such unusual wall formation, or that different patterns of wall formation actually occur in different plant groups.

There are disagreements and perhaps some confusion about endosperm wall formation because the early wall stages are tenuous, but probably mainly because there are four different sequentially produced walls, some with no counterparts in other tissues (Olsen et al., 1995). These wall configurations will not be considered further here, but their existence emphasizes that the transition from coenocyte to multicellular phase involves unusual, and quite probably unique, cell wall formation. Brown et al. (2002) provides the most recent and detailed review of the intricacies of endosperm wall formation. After endosperm cells have been formed they can continue to divide by ordinary cellular mitosis and cytokinesis, but later cell divisions become inhibited because endosperm nuclei gradually reach certain high levels of DNA proliferation or because abnormal amitotic events occur.

Examples of the coenocytic/multicellular type of endosperm development among dicots are lettuce and sugar beet (*Beta vulgaris*, Chenopodiaceae); the latter is illustrated in this chapter by combining the general views of successive stages shown in Jassem (1973) with selected detailed views from an earlier study by Artschwager (1927). Sugar beet is also of interest because its endosperm is eventually replaced by perisperm, described later in this chapter. The first PEC nuclear division in sugar beet occurs 7–10 hours after pollination, followed by synchronous mitoses to produce four PEC nuclei as the chalazal end of the embryo sac elongates and expands, which Artschwager (1927) called the "caecum" (a blind pouch).

By the 8-nucleate PEC stage, a small embryo has developed. Several synchronous waves of mitosis sweep through the population of PEC nuclei following the 32-nucleate stage, which produces a large number of nuclei by 7 days after pollination, and at the same time several large vacuoles have formed in the PEC. Shortly after, cell walls begin to form around PEC nuclei and their associated cytoplasm, starting next to the embryo, which is now in the globe stage. The large central vacuole has pushed the cytoplasm with its many nuclei to the cell periphery. The embryo sac has been gradually elongating and curving around the proliferating nucellar tissue, which will become the perisperm. About 10 days after pollination, the embryo has just initiated cotyledons and the PEC has extended its multicellular phase halfway to the chalazal end of the embryo sac.

As the last few endosperm cells form, the seed has already attained its full size, and the embryo sac curves around the mature perisperm. A more detailed view at this stage shows a small cluster of PEC nuclei remaining at the chalazal end of the embryo sac, which curves around the large-celled perisperm. The endosperm finally becomes completely multicellular by about 14 days after pollination; from then until about the 18th day, endosperm cells are gradually absorbed into the growing embryo. The seed is mature shortly after day 18. Changes in the perisperm are undescribed.

Maize provides a representative example of coenocytic/multicellular endosperm in a monocot, as described by Randolph (1936), with supplementary information from Kowles and Phillips (1988). The PEC begins its coenocytic phase by a nuclear division 26 hours after pollination, which is only 2–4 hours after the central cell was fertilized.

By 34 hours post-pollination, the PEC has 8 nuclei but the zygote is still undivided. The PEC nuclei divide in synchrony, advancing at intervals through 2–4–8–16–32–64–128 levels. By the 3rd day, nuclei are dividing synchronously to reach the 256-nuclei level, and the nowgrowing embryo proper has four small cells and a large suspensor.

The PEC begins to subdivide itself into cells at about this time. Cell walls start forming adjacent to the young embryo during the 4th day after pollination. Wall formation moves rapidly in a wave toward the chalazal end of the embryo sac, and by the end of the 4th day, the former PEC is entirely subdivided into endosperm cells, each with a single nucleus.

Subsequently, by what seems to be typical mitosis and cytokinesis, more endosperm cells are produced. This activity reaches a peak about 8 days after pollination; then the rate of cell division declines to a very low rate by day 12 (Kowles and Phillips, 1988). Only the peripheral endosperm cells continue to divide in a very uniform manner, similar to the cambium layer in the stem of a woody plant, an activity that produces the regularly arranged specialized outermost aleurone cell layer (Randolph, 1936).

The total number of endosperm cells formed in a maize kernel has been calculated on the basis of number of nuclei per total endosperm tissue. The number ranges from 176,000 to 880,000, according to different investigators. A similar wide range of cell numbers has been reported for wheat. Estimates of the total number of endosperm cells in barley range from 170,000–270,000 (Kvaale and Olsen, 1986). Such wide-ranging estimates indicate either the difficulty of calculating the true number of endosperm cells or that the number of endosperm cells is in fact quite variable.

As cell division subsides in the maize endosperm, however, cell nuclei increase in average volume and DNA content, and both reach their peak 16–18 days after pollination. Not all endosperm nuclei are affected equally, and at later stages of kernel development there is a heterogeneous population of larger and smaller endosperm cells, with ploidy levels ranging from 3C to as high as 690C. It is probable that as nuclei become larger because of increasing DNA levels they reach a level at which they are unable to undergo normal mitosis. Although there is great individual variation among cells of the maize endosperm, the trend in development as a whole is quite regular: cell division ceases gradually from kernel base to tip, whereas starch and protein accumulation in endosperm cells proceeds from tip toward base. Kowles and Phillips (1988) concluded that DNA amplification multiplies gene expression and therefore enhances the ability of endosperm cells to accumulate storage products. Briarty et al. (1979) described the development of coenocytic/multicellular endosperm in wheat by analyzing light and electron micrographs taken at 11 developmental intervals between 2 and 52 days after anthesis.

The initially coenocytic PEC starts to become multicellular during days 2–4. The PEC does not increase its volume during this time because it is expending energy subdividing itself into cells instead of growing larger. But after the PEC becomes completely cellular, total endosperm volume begins to increase and more endosperm cells are produced by mitosis, finally reaching the impressive number of 100,000 cells by day 16. Endosperm volume hesitates briefly and then continues to increase as the cells grow in size, but not in number, until the 35th day. Endosperm volume then decreases from day 35 to the mature grain at day 52 because the cells lose water gradually. There is an opposite trend of steady increase in starch content up to day 36, at which time starch occupies 65% of the volume of each endosperm cell.

A second investigation of wheat endosperm by Chojecki et al. (1986) added two points of interest. First, they provided evidence from their own study and from the literature that the final number of endosperm cells is more than double that reported by Briarty et al. (1979). Second, they detected DNA increases in endosperm cells without accompanying cell division, resulting in a population of cells with the initial 3C levels and others with 6C, 12C, and 24C. They cited other studies of wheat, as well as studies on pea and vetch, in which similar increases in endosperm cell DNA were reported. They did not, however, discuss the possible significance of such DNA replication for wheat endosperm.

Another monocot with coenocytic/multicellular endosperm is zephyr lily (*Zephyranthes lancasteri*), a cultivated relative of amaryllis. The coenocytic phase switches to the multicellular phase 7 days after pollination, and cells are produced thereafter until seed maturity at 18 days.

The most spectacular examples of coenocytic/ multicellular endosperm are from coconut palms and their relatives, which have seeds as large as canteloupes. Their protracted coenocytic phase is known familiarly as coconut milk, a potable mass of protoplasm with hundreds of thousands of nuclei. The later multicellular phase of endosperm is the edible coconut meat.

HELOBIAL ENDOSPERM

The helobial endosperm is really a variation of the coenocytic/multicellular pattern. An initial cell division of the PEC produces two cells, of which the micropylar cell next to the zygote is the largest. It becomes coenocytic, and then multicellular; therefore, it seems to behave identically to the coenocytic/multicellular pattern. The smaller chalazal cell also becomes coenocytic, with few to many nuclei depending on the species, but it usually does not subdivide itself into cells. This way of forming endosperm is mostly restricted to monocots. Davis (1966) reported it from 17 families, 14 of them monocots. The term "Helobial" refers to Helobiales, the Order of monocots in which this type of endosperm was first discovered, and in which it is common.

COENOCYTIC ENDOSPERM

In the coenocytic endosperm type, the PEC proliferates nuclei, but the coenocytic phase is never followed by subdivision into cells, thus the PEC remains as a coenocyte until seed maturity. Such an extreme form of endosperm is uncommon, but it has been reported for silver maple, *Acer saccharinum* (Haskell and Postlethwait, 1971), other maple species (Davis, 1966), and broad bean, *Vicia faba* (Leguminosae) (Kapoor and Tandon, 1964), among several other legumes. Many other legumes delay cell formation for such a long time that the PEC remains coenocytic for most of the life span of the endosperm

(Davis, 1966). Many grasses also have so-called "liquid endosperm" (Terrell, 1971) at seed maturity, which can persist for decades in seeds from dried herbarium specimens.

A well-illustrated example of true coenocytic endosperm in a cultivated plant was not found, but the peanut (*Arachis hypogaea*) is among the many legumes that almost qualify. Prakash (1960) studied whole embryo sacs dissected out of ovules at different stages in addition to sectioned ovules, an unusual but useful combination of procedures for endosperm studies. The peanut PEC is full of starch before and during fertilization. After fertilization the coenocytic PEC remains rich in cytoplasm, but starch decreases during the first few nuclear divisions.

The zygote divides when the PEC has 2–4 nuclei. By the time the PEC has eight nuclei all starch has disappeared and the young embryo has four cells. The embryo continues to grow, and by the late globe stage it impinges on the coenocytic PEC, which is now filled with hundreds of nuclei. A detailed look at the embryo and adjacent area at about this stage shows some of the numerous and synchronously dividing nuclei. Later stages are not shown here, but following initiation of cotyledons the coenocytic PEC forms a central vacuole surrounded by peripheral cytoplasm containing a large number of nuclei. The developing cotyledons absorb nutrients from the coenocytic PEC, which shows a corresponding decrease in volume. Only very late in embryo development does the now greatly diminished PEC subdivide itself to form two thin layers of peripheral endosperm cells, which persist to seed maturity.

ENDOSPERM HAUSTORIA

A haustorium (Latin for "something that drinks or sucks") is a cellular structure that grows into or around another structure and absorbs water and nutrients from it. An endosperm haustorium is typically a simple or branched tubular sac, usually coenocytic, which penetrates (probably intercellularly, but published accounts usually do not tell) the surrounding nucellar tissue.

Such diverticulata have been recorded for at least 73 families among both dicots and monocots (Davis, 1966), and they can occur in species with any of the four types of endosperm. Many variations have been described and illustrated, for example in Maheshwari (1950), Chopra and Sachar (1963), and Vijayraghavan and Prabhakar (1984). Endosperm haustoria are common in certain families with cultivated plants, such as Fabaceae and Cucurbitaceae; which illustrates the long, cordlike chalazal endosperm haustorium in the cucumber, *Cucumis sativum*. Haustoria described from some cultivated legumes are, in contrast, typically just small coenocytic pouches at the chalazal end of the endosperm.

Because they are often branched and deeply embedded in the nucellus, it has been speculated that haustoria draw additional nutrients into the endosperm. This is a reasonable speculation, but there is no direct evidence to support this or any other hypothesis regarding their function.

PERISPERM

Perisperm consists of nucellar cells; therefore, it is composed of ordinary diploid parental sporophyte cells. Perisperm is presumed to provide nutrients to either the endosperm or the embryo, or both. Why it forms, and what advantage it might have over endosperm is unknown. The standard sources (Davis, 1966; Maheshwari, 1950; Chopra and Sachar, 1963; Jacobsen, 1984; Vijayraghavan and Prabhakar, 1984) say little about it, which indicates that not much is known about it.

Perisperm has been reported mostly from members of a group of dicot families collectively known as the Centrospermae, which refers to the mature embryo curved around the nucellar swelling that is perisperm. Sugar beet is a good example from a cultivated species. Perisperm is also known from some monocots; the agaves are a familiar group in which it occurs.

A detailed study by Prego et al. (1988) of the seed crop plant quinoa (*Chenopodium quinoa*, dicot family Chenopodiaceae) showed that endosperm in the mature seed is only a remnant, 1–2 cell layers thick, whereas the perisperm is a spheroidal central mass of dead, thin-walled cells filled with angular starch grains. Endosperm and embryo, in contrast, have quite different storage components: lipid and protein bodies, globoid crystals of phytin, and proplastids with phytoferritin. Perisperm in this species is distinctive anatomically and physiologically, and other species may have similar perisperm, but it is a tissue that has not been studied very much.

MOVEMENT OF CARBOHYDRATES INTO ENDOSPERM

Nutrients can enter the embryo sac and endosperm from any surrounding tissue, although the chalaza seems the likeliest site since that is where the ovular vascular bundle ends in most plants. Transfer cell wall ingrowths have been reported from the periphery of the embryo sac in bean (Yeung and Cavey, 1988), sunflower (Newcomb, 1973), and several other plants (Vijayraghavan and Prabhakar, 1984). Such wall ingrowths greatly expand the cell membrane, which is thought to mediate transfer of nutritious solutes more efficiently to the developing endosperm and embryo.

Various anatomical modifications in grasses have been suggested to help move carbohydrates to the developing endosperm. In wheat, for example, Zee and O'Brien (1971a) described conspicuous transfer cells located in the node where the caryopsis is attached to the rachis. In at least a half-dozen species of grasses the caryopsis has been found to have a placental pad, a local area of seed coat and aleurone layer (outermost layer of endosperm) at the base of the caryopsis, adjacent to the ovular vascular bundle on the opposite side from the embryo. This anatomically modified zone is probably an entryway for carbohydrates to pass into the developing endosperm.

The simplest placental pad is merely an elongation and/or proliferation of transfer cell wall thickenings in the aleurone cells in a local oval or circular area, a modification that increases the cell membrane and aids short distance transport. Rost and Lersten (1970) showed such a modification in the weed yellow foxtail grass (*Setaria lutescens*). In Japanese millet (*Echinochloa utilis*) a swollen parenchymatous pad of elongate aleurone cells with wall ingrowths occurs in the same location in the grain, and was termed a "nucellar projection" by Zee and O'Brien (1971b). They interpreted it as a "structural bottleneck," a modification to increase entry of solutes into the ovary. A similar nucellar projection in rice occurs farther up the dorsal side of the grain, but it is still intercalated between the vascular bundle and the aleurone layer, which is 4–5 cells thick in this area. Hoshikawa (1973, 1984b) regarded the nucellar projection in the rice grain as a primary entry point for nutrients into the endosperm.

A more elaborate adaptation occurs in sorghum (*Sorghum bicolor*). Maness and McBee (1986) provided both anatomical and physiological evidence for the entry of carbohydrates into the caryopsis via the "placental sac," a cavity within the nucellar parenchyma located between the ovular vascular bundle and the aleurone layer. By about 15 days after pollination the sac is filled with an opaque fluid rich in glucose and fructose, with smaller amounts of sucrose. The adjacent aleurone cells have well-developed transfer cell walls, which increases membrane surface area for efficient short distance transport. The placental sac and surrounding area starts degenerating at about 20 days post-pollination and continues to do so until grain maturity. Maness and McBee interpreted the placental sac as a temporary reservoir for sugars awaiting transport through the modified aleurone cells and into the endosperm during its period of most active growth.

STORAGE PRODUCTS IN ENDOSPERM

Starch is the most common and abundant component found in endosperm consisting of dead cells at maturity, as in grasses. The extensive studies on carbohydrates in cereal endosperm were reviewed by Duffs and Cochrane (1982). Sucrose is the entering sugar, which is then converted into starch by a complex and still incompletely known series of reactions. Starch accumulates in wheat at a steady pace throughout grain development, which also seems to be true in general for cereals. In particular, the amylose content of starch increases until, at grain maturity, the starch consists of about 25% amylose and

75% amylopectin. In contrast to plants with typical starchy dead endosperm cells, plants with living mature endosperm cells tend instead to have mostly carbohydrates, lipids, and proteins (Jacobsen, 1984). Carbohydrates of some species are stored mostly as polysaccharides in modified endosperm cell walls. In certain palms (DeMason et al., 1983; DeMason, 1986) the extremely thick walls are so hard that the endosperm is sold commercially as "vegetable ivory." Coffee (*Coffea arabica*, dicot family Rubiaceae) also has thick, mannose-rich endosperm cell walls, which are digested later by the young seedling. Such thick-walled endosperm probably also has a secondary function of protecting the embryo in the mature seed.

Lipids are common in endosperm, but their relative abundance varies among plants. Lipids in most oil seeds are storage products in the cotyledons and therefore derived indirectly by transfer and transformation from endosperm. Palms, especially the African oil palm (Fernandino et al., 1985), are conspicuous exceptions because their endosperm is extremely oil-rich.

Protein is also found in endosperm, stored in discrete protein bodies in both dicots and monocots (Lott, 1981). In cereal grasses the amount of endosperm protein ranges from 10–17% of dry weight of the grain in wheat and oats to a somewhat lower level of about 6% in maize and rice (Payne, 1986). A well-studied type of protein body is the so-called "aleurone grain" of aleurone cells (see next section). In cereals, these protein bodies are abundant in the aleurone layer, but their numbers decline markedly in the first few adjacent endosperm cell layers. The chemical composition of storage proteins is complex (Payne, 1986; Spencer and Higgins, 1982) and cannot be dealt with further here. The fact that 70% of the protein eaten by humans and farm animals is from seeds attests to the importance of this endosperm component.

ALEURONE LAYER AND MATURE ENDOSPERM

Aleurone (Greek, meaning approximately "flour") is usually the outermost single cell layer of endosperm. These thick-walled living cells contain characteristic protein bodies called aleurone grains. An aleurone layer occurs among both monocots and dicots, but it is best known from grasses. The aleurone layer in maize results from the activity of a cambium-like meristem on the periphery of the endosperm that produces new cells during the period 12–16 days after pollination (Randolph, 1936). The outermost cell layer that is formed gradually stops dividing during day 18–22, and thereafter it assumes some of the characteristics of an epidermal layer even though it is not exposed to the outside. The aleurone protein bodies form within vacuoles, in contrast to the protein bodies of nearby endosperm cells, which are initiated by protrusions of rough endoplasmic reticulum (Kyle and Styles, 1977). The aleurone layer in many, perhaps most, grasses is a single layer of living cells (Figs. 9.9B, 9.10). In barley, however, it is 3–4 cell layers thick. Among cereal grasses that have been investigated, the starchfilled central endosperm is composed instead of dead cells packed with starch grains.

An important function of the aleurone layer in cereals is the release of enzymes that digest starch in the central endosperm cells. When the embryo becomes rehydrated at the time of germination, gibberellins are released and diffuse out to the aleurone layer, stimulating it to produce and release starch-digesting hydrolytic enzymes, which begin to convert endosperm starch to sugars for seedling growth (Jacobsen, 1984).

A specialized cell layer at the periphery of the endosperm also occurs in some plants other than grasses. A layer similar to that of the aleurone of grasses was described long ago by Stevens (1912) in buckwheat (*Fagopyrum esculentum*, Polygonaceae), and a convincing example of an aleurone layer was shown more recently by Groot and Van Caeseele (1993) in the dicot oil plant, canola (*Brassica napus*). The best known example of an aleurone layer among dicots is probably that of fenugreek (*Trigonella foenum-graecum*), a legume that has abundant but dead endosperm cells at maturity. The aleurone layer storage product consists almost exclusively of galactomannan, a hydrophilic polysaccharide deposited in such large quantities between cell membrane and cell wall that it occludes the cells. There is some evidence that the aleurone layer of fenugreek is involved in galactomannan digestion (Reid and Bewley, 1979).

Another legume with galactomannan-rich endosperm is guar (*Cyamopsis tetragonoloba*), in which the aleurone layer is several cells thick. The abundant endosperm of another legume, the carob bean, *Ceratonia siliqua,* is also rich in galactomannans, but here the endosperm cells are alive and there is no aleurone layer (Reid, 1985). The endosperm galactomannans of these legumes absorb and retain water, which keeps the embryo from drying out during germination. The galactomannans are absorbed later by the seedling (Reid and Bewley, 1979). The property of water absorption has been exploited by selling purified galac tomannan in the form of diet pills, which can absorb water and swell in the stomach, giving a feeling of fullness.

The so-called aleurone layer in mustard seeds has been shown by Bergfeld and Schopfer (1986) to be derived from a specific layer of the inner integument instead of from the endosperm. At seed maturity, however, it lies next to the endosperm, and upon germination this layer performs the same functions as an aleurone layer. They suggested that other dicots reported to have an aleurone layer need careful study to determine whether it is really a product of the endosperm or is instead part of the seed coat.

The thick-walled peripheral two cell layers of rather specialized endosperm that remain in the mature seed of lettuce has not been called an aleurone layer (Jones, 1974; Psaras, 1984), but the function of these two layers is clearly not to provide nutrition. This aleurone-like layer seems instead to restrain embryo growth, thus helping maintain seed dormancy. Jones (1974) provided evidence that when the seed germinates, this layer secretes enzymes that soften its own walls, allowing the embryo to resume growth and push its way out of the seed.

The presence of an aleurone layer, or any modified outer endosperm layer similar to it, is not a predictable feature of plants, even of those with copious endosperm. There is no aleurone layer in castor bean (*Ricinus communis*, Euphorbiaceae), for example, but the endosperm cells themselves are living and become active upon germination (Vigil, 1970). DeMason (1986) and DeMason et al. (1983) showed, as another example, that there is nothing corresponding to an aleurone layer in date palm or California fan palm. They pointed out that there are few studies of mature endosperm outside the cereal grasses; therefore, knowledge of the occurrence and distribution of the aleurone layer in angiosperm seeds is sparse and consists of scattered observations. Even based on this incomplete knowledge, one can conclude that the outermost endosperm cell layer or layers, like the main body of the endosperm, has evolved different specializations in different groups of flowering plants.

FUNCTIONS OF ENDOSPERM

Endosperm is ephemeral even though it persists well into the seedling stage in many species, particularly among monocots. Because it is short-lived, similar to the suspensor of the proembryo, endosperm has been able to undergo modifications and take on various functions without affecting embryo or seedling structure. A few of these functions were mentioned earlier in this chapter. Endosperm functions that have received some verification, as well as more speculative ones, are discussed here.

Nutrition for the developing embryo is the primary function of endosperm, and evidence for its partial or complete absorption by the embryo. Evidence that endosperm provides nutrition for seedlings comes from direct observation of haustorial cotyledons in monocot embryos, and the disappearance of endosperm as seedlings grow.

Before it contributes nutrition to the embryo, however, at least the early part of the coenocytic PEC stage seems to be devoted mostly or entirely to its own internal growth. The best evidence for this is from Bennett et al. (1975), who showed that nuclei in the wheat PEC proliferate at a rapid rate during the coenocytic phase, probably because nutrients enter through the chalazal cluster of polyploid antipodal cells, whose high DNA levels enable them to attract nutrients. The extended period of nuclear or cellular proliferation before the zygote divides is also indirect evidence of a preliminary period of internal PEC growth.

A second function of endosperm in many species is to impose dormancy on embryos, either by chemical or physical means. There is evidence that at later stages of embryo growth in some plants the concentration of growth inhibitors, such as abscisic acid, increases in the endosperm (Walbott, 1978). Thick endosperm cell walls of many plants, as described earlier, probably impose dormancy by physically sealing off the embryo from gas exchange. A well-known example of this is lettuce, described in the previous section of this chapter.

A third function can arise from the same physical features that impose dormancy. Thick endosperm cell walls may also help protect the dormant embryo from damage and prevent germination from occurring except when growing conditions are good. This also helps spread seed germination over a long period of time for the benefit of the species. In this way the endosperm in some species appears to serve the same function as a hard seed coat, or at least it can supplement the seed coat.

A fourth endosperm function is its active involvement in germination of some species, which has been studied intensively in cereal grasses. The grass aleurone layer has been shown to be a site of enzyme production and secretion, although the scutellum (the specialized grass cotyledon) also contributes a portion of this (Jacobsen, 1984; Jones, 1985). Plants that lack an aleurone layer but have living endosperm cells possibly have some analogous mechanism by which endosperm can aid germination. A fifth function is also associated with germination. Endosperm in some species absorbs and retains water to keep the seedling moist.

There is good evidence that the hydrophilic galactomannans of some legumes with abundant endosperm, as described earlier, hold water tenaciously for the benefit of the germinating seedling. Reid also cited earlier work describing mucilaginous endosperm in other species—i.e., endosperm capable of absorbing large amounts of water. This capability is common enough that it even convinced some early investigators that water absorption and retention was the main purpose of endosperm. The ephemeral endosperm is foremost a nutritive tissue, but it has evolved to serve other functions as well in different plants. These functions may operate at any stage between fertilization and seedling. As a nutritional tissue, endosperm runs the gamut from nonexistent in most orchids and a few other families to a massive store of carbohydrates in coconut and related palms.

SPECULATIONS ON ENDOSPERM VARIATION

The coenocytic/multicellular type of endosperm, which is most common, can be viewed as an efficient two-step adaptation. During the early coenocytic phase, nutrients flow rapidly and easily into it without investment in cell walls. At a later stage, as progressive enzymatic and physical destruction and absorption of the endosperm occurs from growth of the embryo, it seems advantageous to have the endosperm divided into individual cells because they can be destroyed and absorbed piecemeal without harm to what might otherwise be a vulnerable single large PEC. The existence of the coenocytic type of endosperm, or at least of a protracted coenocytic stage before walls form late in development (e.g., peanut described earlier), could be explained by postulating that transfer of nutrients to the embryo is so rapid in these species that the PEC is only a transitory intermediary. Different enzymes are probably involved, which enable nutrients to be absorbed from the PEC coenocyte across its cell membrane over a long period without destroying it.

Early-acting enzymes must be capable of mediating reactions that move nutrients out of the PEC and into the embryo without damaging the cell membrane, which must be kept intact to allow the PEC to function as a living cell. Endosperm in which cells form within the PEC only around the periphery of the developing embryo could be in an intermediate position of balancing between the advantages of rapid influx of nutrients into a coenocyte and the pressures, both enzymatic and physical, of the growing embryo impinging on the growing endosperm.

The occurrence of the multicellular type of endosperm in many families means that there is no absolute need for an early coenocytic stage; it is possible that in species with this type the embryo grows more slowly than it would with a coenocytic PEC phase. It is also possible that enzymes capable of extracting nutrients from a coenocytic PEC are not present (did not evolve) in species with this type.

The helobial type of endosperm is the most difficult type to speculate about. It really combines two types, and therefore the speculations made earlier do not seem to apply here in any logical manner. A plausible suggestion is that the chalazal coenocytic cell is possibly a variant form of endosperm haustorium that does not aggressively invade surrounding tissue but retains a coenocytic configuration to more easily transfer substances to the micropylar multicellular endosperm.

The variations in endosperm structure and behavior described in this chapter raise the same problem that will be raised in the next chapter concerning the embryo suspensor, namely, that variations among these structures make it very difficult to formulate a common generalization that applies to all of them.

10

The Embryo

The embryo sac (female gametophyte), unlike the briefly independent pollen grain (male gametophyte), always remains embedded in the ovule. The embryo likewise remains nurtured within the embryo sac. This results in a remarkable nesting of generations: new sporophyte (embryo) within female gametophyte (embryo sac) within old sporophyte (ovule). In addition the fertilized central cell produces endosperm, which becomes still a different enveloping polyploid layer around the embryo. This layering of generations means that the seed is a genetically diverse end product of reproduction, without even considering its enveloping fruit.

INTRODUCING THE COTYLEDON(S)

Before starting with the zygote, a brief discussion about the peculiar first appendage produced by the embryo, the cotyledon. The origin of the term (Greek, meaning "cup-shaped hollow") is obscure. Perhaps the name was given because cotyledons of many dicot seedlings gradually wither and turn up around the edges as their stored food is utilized, which gives them a somewhat cuplike appearance.

The cotyledon of the mature embryo has the unexpected distinction of providing the primary taxonomic character that delimits and names the two large angiosperm groups; Monocotyledoneae (monocots) have one and Dicotyledoneae (dicots) have two. Why should this difference exist almost without exception? Among monocots two common features probably contribute:

- The sheathlike single cotyledon completely encircles the shoot apex, leaving no space for a second cotyledon.
- The monocot cotyledon typically remains in the seed as a haustorium, enlarging as it absorbs endosperm upon germination for translocation to the seedling, and leaving little room in the seed for a second cotyledon.

Why do dicots have two cotyledons (a few do have three or more)? No sure answer exists, but one can speculate that a space limitation around the tiny embryonic shoot apex is a factor, and also that there is only enough room for two cotyledons to reach mature size within the seed. Most dicot cotyledons either absorb endosperm during embryo development, as in peas and peanuts, or they absorb less endosperm and instead expand after seed germination and become photosynthetic, as in sunflowers and many other dicots.

THE ZYGOTE

The sperm nucleus that merges with the egg nucleus (technically called karyogamy—"marriage of nuclei") contributes a second set of chromosomes, which restores the 2N sporophytic genetic constitution and transforms the egg cell into the zygote (Greek, meaning "yoked together"), the first cell of the embryo. The zygote of most species does not divide immediately after fertilization; instead, it waits until the PEC has produced some endosperm. A reasonable speculation is that a certain amount of PEC activity is needed to provide some nutritional, hormonal, or other stimulation.

During the so-called "resting period" before it divides, largely unknown subcellular events occur, but overall the zygote may stay the same size, as reported for example in barley (*Hordeum vulgare*) (Norstog, 1972), or it may shrink to half of its original size before dividing, as in cotton (*Gossypium hirsutum*) (Pollock and Jensen, 1964) and normal selfed *Hibiscus* (Ashley, 1972) (both dicot family Malvaceae).

It seems more logical that a zygote would enlarge before it divides, and many investigators have reported this, for example Wojciechowska and Lange (1977) in wheat (*Triticum aestivum*) and Takeuchi (1956) in kidney bean (*Phaseolus vulgaris*). Natesh and Rau (1984) mentioned several other examples of zygotes that either swell or shrink. The mustard family (Brassicaceae) includes many examples of zygotes that enlarge by elongating to become conspicuous unicellular filaments 10–15 times longer than wide before they divide. The mustard plant (*Brassica campestris*) forms such a wormlike zygote.

Few investigators mention changes in zygote size or shape, or speculate about why such changes should occur, but it seems likely that they involve osmotic relations, so that water moves in or out as controlled by the zygote itself or in response to some influence of the PEC, which has invariably already started developing endosperm around it.

The zygote must rearrange its internal components (establish polarity) in order to locate where the first wall will occur after mitosis, an important if not critical event in determining the disparate fates of the two resulting cells. Knowledge gained from studies of polarity in free-living zygotes, especially in the marine green alga *Fucus*, may also apply to angiosperms.

The spot on the egg where the *Fucus* sperm enters, supplemented by cues from light and gravity gradients, determines the future germling axis. Just before the zygote germinates, Golgi-derived vesicles accumulate at what will become the basal end, which fixes its polarity, and these vesicles deposit new wall material and initiate growth here. Microtubules extend from the base to the nucleus, which orients spindle fibers perpendicular to the microtubules and orients the position of the first cell wall. Although Scheres and Benfey (1999) suggested in their review paper on asymmetric cell division that a similar mechanism might occur in angiosperms, they admitted "...much of the control of the first zygotic cell division in higher plants remains unknown."

Light and shade cannot influence zygote polarity in angiosperms, but proximity to the micropyle has been suggested as a factor. Contrary evidence, however, comes from zygotes of some species that have been seen occasionally in a lateral position in the embryo sac, away from the normal position adjacent to the micropyle.

Such zygotes divide nevertheless and form a normal suspensor initial that contacts the embryo sac wall at that point, while the apical cell develops into an embryo that grows laterally or obliquely across the central cell. This has been shown by Haskell and Postlethwait (1971) in silver maple (*Acer saccharinum*) and reported by Mestre and Guignard (1973) from several other species. Such examples seem to contradict the argument that a specific orientation of the zygote is required to determine its polarity.

Changes within the zygote, such as vacuole size, movements of various organelles, and increases in various substances, are probably related to polarization, but descriptions of such changes do not yet collectively provide a coherent explanation. Raghavan (1986) reviewed published reports and concluded "No generalizations can be made about the major ultrastructural upheavals that occur in the egg after fertilization."

A similar conclusion was reached by Yadegari and Goldberg (1997), although they did cite some studies that reported involvement of the subcellular microtubule framework in establishing polarity. In a recent review that mentioned zygote polarity, Aida and Tasaka (2002) concluded "...little is known about how the initial polarity is generated." They did, however, suggest that directional vesicle transport and polar auxin transport are probably involved, and they also speculated that "maternal information" could in some way provide a "positional clue." What stirs the zygote to its profoundly important first division still remains mysterious.

Several studies have shown that the egg cell either lacks a cell wall or has only a partial (patchy) wall at time of fertilization. This necessary omission allows the sperm to contact and penetrate the naked egg cell membrane. But the resulting zygote seems to require a complete cell wall before it divides, which has been reported for several species examined by electron microscopy. For example, Mogensen and Suthar (1979) showed that at the time of fertilization most of the distal (chalazal) part of the tobacco egg cell lacks a cell wall, but that by 40–50 hours after fertilization the zygote wall is complete.

Callose occurs around the zygote, at least in some species. Williams et al. (1984) demonstrated that several species of *Rhododendron* (Ericaceae) retained callose at least until after the zygote divides at some time during the first two days after fertilization. Callose was not detected around the egg cell in unfertilized ovules of flowers prevented from being pollinated; therefore, fertilization seems to be required before callose is secreted.

They implied that callose around the zygote has also been seen but not identified as callose by a few previous investigators. Williams et al. (1984) felt that this callose, like that secreted around microspore- and megaspore-mother cells, is a barrier needed to isolate the sporophytic zygote temporarily from the surrounding gametophyte.

A novel interpretation of the zygote resulted from the observation by Bruck and Walker (1985a,b) that a cuticle forms on the zygote of *Citrus jambhiri*, a lemon species. They proposed that the zygote is the first epidermal cell, and that it produces more epidermal cells by anticlinal divisions and generates internal cells by periclinal divisions. This citrus species, like some others, produces extra vegetative embryos (a process called polyembryony—see section at end of this chapter), so Bruck and Walker were able to examine one to six additional asexual zygotes in addition to the primary zygote.

They reported that a cuticle forms on both sexual and asexual zygotes, but did not mention a cuticle on the first two daughter cells; they did, however, report a cuticle from older embryos but not from the suspensor. Bruck and Walker are among only a very few workers who have considered the question of when a cuticle forms on the embryo, which is important because the cuticle can be considered a barrier that prevents, or at least seriously restricts, movement of substances in or out. On theoretical grounds one could therefore doubt that a zygote would have a cuticle, but little information exists on this for any stage of embryo growth. The subject of cuticle formation will be taken up again later in the section on nutrition of the embryo.

PROEMBRYO INITIATION

Embryo growth begins when the zygote divides, which often takes some time to occur. It includes elapsed time between fertilization and zygote division for some representative cultivated plants, as well as how much endosperm has proliferated when the zygote does divide. The zygote may divide soon after fertilization, as in lettuce, but more commonly it waits for various longer periods, even up to several weeks, as in pecan and coffee. The zygote seems to divide earlier in herbaceous plants than in woody plants, but whenever it divides there is already slightly to very welldeveloped endosperm, which implies that the zygote requires some nutrition or other stimulative substance from either the PEC coenocyte or the multicellular endosperm before it can divide.

Embryo development beyond the zygote is more or less continuous, but stages are identified for more precise description. The proembryo stage extends from zygote division through the spherical globe stage, and ends when cotyledons are initiated in dicots, or when the shoot apex is initiated in monocots. Zygote cytokinesis produces a transverse or obliquely transverse wall (rarely a sagittal wall). The apical cell is usually much smaller, and it abuts the PEC; the usually much larger basal cell faces the micropyle.

Sivaramakrishna (1978) reported that these are not always the proportions, however, after he measured 2-celled proembryos of 100 species from illustrations in published papers. In 34 species both cells were either equal in size or the apical cell was actually larger. Although the relative sizes of the daughter cells may not be invariable, the orientation of the apical cell toward the chalaza, which is called endoscopic (Greek for "looking inward"), is evidently of critical physiological importance. The endoscopic apical cell and cells derived from it often produce all of the cells that will comprise the mature embryo, and the basal cell and all cells derived from it will produce only an ephemeral structure called the suspensor (described in the next section). In other species the boundary between proembryo and suspensor is not absolute, and at least some cells produced along this border can become part of either entity, as in *Cocos* and maize described later.

The basal cell in *Arabidopsis thaliana* (Brassicaceae) has even been shown to have the potential to produce an embryo instead of a suspensor, although it is normally prevented from doing so by an as yet unknown mechanism controlled by the apical cell and later the proembryo. Schwartz et al. (1997) reviewed this and other molecular genetic studies on suspensors. It is possible that the fate of the basal cell is controlled by the apical cell, but the question remains as to why the two cells that result from zygote mitosis each initiate completely different entities. Different cell components resulting from zygote polarity are probable factors, but at a different level the answer must include a consideration of what is necessary for subsequent development of the plant.

The embryo will grow into a seedling and eventually become an adult plant. Its architectural form and pattern of growth must be established in the embryo, which means at embryo inception— which means in the apical cell. The ephemeral suspensor, in contrast, functions only during early embryo development, typically only up to cotyledon and root meristem initiation, and in many plants it is only a rudimentary nubbin or even degenerates after just one or only a few cells have been produced. In other words, remarkable latitude is tolerated for the basal cell and its progeny, but little or none for the apical cell and its progeny because they must establish the morphological pattern for the embryo that will be continued in the adult plant.

THE SUSPENSOR

The suspensor is ephemeral and extremely variable in form. Insight into why such an entity should even exist may be gained by specu lating. Imagine that a zygote drops directly to the ground instead of being nurtured in the ovule. Just as a seedling must quickly establish a root system, this imaginary zygote must divide unevenly into a small apical cell and a large basal cell, because the latter cell needs more energy to elongate rapidly into a simple absorbing rhizoid system to convey water and dissolved substances. The small apical cell can then begin to photosynthesize and generate more cells to organize a shoot system and a true root system to replace the temporary rhizoid system. In other words, in this imaginary scenario the suspensor is a temporary root substitute.

The real zygote germinates (divides) within the embryo sac, where it is always in contact with either the nutrient-rich PEC coenocyte or multicellular endosperm. As in the imaginary scenario, however, the basal cell and its progeny may have one or more temporary functions, but not necessarily the same ones as in the imaginary scenario. In the confined embryo sac, with sustenance at hand, a precisely organized suspensor seems unnecessary, as attested by its remarkably diverse forms among angiosperms. Suspensors may be filamentous or bulky, unbranched or branched, and some even invade haustoria-like into adjacent nucellar tissue. Still other suspensors are reduced to just a few cells—e.g., the rudimentary two- or three-celled suspensor of cotton (Pollock and Jensen, 1964)—or they might degenerate completely after only 2–3 cells form, as in silver maple (Haskell and Postlethwait, 1971). In some species the suspensor consists of only the original basal cell.

Wardlaw (1955) surveyed published descriptions of suspensors, and concluded "...we can find practically every gradation between species with a well developed suspensor and those in which it is inconspicuous or virtually absent." Such extreme suspensor variability demonstrates that suspensors and their functions can hardly be said to be under rigid control among angiosperms as a whole, and that in many species the proembryo itself must perform most or all of any suspensor functions. Embryos of legumes (Fabaceae) exhibit perhaps the most remarkable suspensors among angiosperm families, ranging from nonexistent to large and/or bizarre forms, as reviewed by Lersten (1983). Especially noteworthy is the unique suspensor of the tribe Vicieae (e.g. broad bean, lentil, chick pea, garden pea), which consists of only four multinucleate cells, two smaller upper and two large lower, each containing 32–64 or more nuclei.

Phaseolus multiflorus, currently called scarlet runner bean (*P. coccineus*), has been a favorite species for investigation because it has a massive suspensor that is relatively easy to remove whole and manipulate for experiments (Yeung and Clutter, 1978). The entire genus *Phaseolus* is remarkable because its species exhibit a range of suspensors from the *P. coccineus* giant to those that are tiny or nonexistent (Nagl, 1974). Nagl also reported that *Phaseolus* suspensor cells exhibit an amazing range of polyploidy, from 8n to an almost astronomical 8,192n! The filamentous suspensor of the mustard family (Brassicaceae) has also received considerable research attention. Such a suspensor from the common weed called shepherd's purse (*Capsella bursa-pastoris*) is illustrated in almost every botany textbook as a representative example of a dicot embryo. Here the field mustard (*Brassica campestris*) is illustrated instead.

The results of many investigations, mostly on the large suspensors of *Phaseolus* and the filamentous suspensors of certain Brassicaceae, were discussed by Raghavan (1986), who concluded the following:

- Some or all cells of many suspensors have wall ingrowths (transfer cell walls).
- The nuclear DNA of suspensor cells may replicate itself to astonishingly high levels, somewhere between 8C and 8,192C.
- Synthesis of RNA and protein in the suspensor is more efficient than in the embryo proper.
- There is considerable evidence that suspensors contain higher concentrations of certain growth hormones than do cells of the embryo proper.

Raghavan's assessment is that suspensors have three functions:

- Actively absorb and transport nutrients.
- Manufacture gene products for the embryo proper.

- Produce growth hormones to regulate early embryo development.

It must be emphasized, however, that the vast majority of species, especially those with small, rudimentary, or nonexistent suspensors remain to be studied.

Observations from several dicot and monocot species, and shown especially well in the carrot embryo (Lackie and Yeung, 1996), show that no cuticle occurs over the suspensor, which probably allows it to absorb substances efficiently for translocation to the proembryo.

Most suspensors appear to become isolated physiologically, and probably physically as well, after radicle (first root) and cotyledon initiation, thereafter degenerating. This suggests that the embryo has then reached a stage of development where it can control its own heterotrophic nutrition until after germination.

The poorly developed embryos of most orchids often have large, branched suspensors (Swamy, 1949). Since little or no endosperm forms in these embryo sacs, such orchids provide indirect evidence that suspensors can absorb nutrients.

A dividing zygote always produces a basal cell that does not become part of the embryo regardless of whether it will produce a functional suspensor. It seems reasonable to speculate that a basal cell is necessary to orient or otherwise prepare the apical cell to initiate the proembryo.

THE EARLY PROEMBRYO PROPER

After the zygote divides, the apical cell produces the embryo, although in some species the basal cell also contributes some cells. The apical cell proliferates first into a small multicellular filament or club-shaped entity, depending on the species. How this early proembryo forms itself into an organized entity has long been considered by many to be significant tax onomically and otherwise. Efforts devoted to describing the supposedly precisely derived tiers of cells in the proembryo have stimulated considerable research on embryo development but they have also, unfortunately, caused this branch of embryology, called embryogeny, to congeal into empty formalism. Because it has received attention in all embryology texts since Coulter and Chamberlain (1903), however, it will be assessed briefly here.

Many 19th century investigators of embryogeny concluded that cell walls in the early proembryo form by precise patterns of cell division that can be classified into types characteristic of certain groups of plants. Other investigators who bothered to sample several proembryos of the same species usually found that these patterns were variable, as Wardlaw (1955) pointed out. Even in 1903 Coulter and Chamberlain had observed that "Undue attention has probably been given to the succession of cell divisions in the earliest stages of the embryo." They concluded that "...the study of a large series of embryos makes it evident that if there is a normal sequence of cell divisions it is being constantly interfered with.

It is probable that when these minor variations are neglected, certain laws of general development will appear that are concerned with the organization of the great body regions rather than with the succession of cell divisions."

Some embryogenists continued to believe that precise patterns exist, that they characterize families or other taxonomic groups, and also that they determine where and how the organs of embryos originate. Formal "laws" that described and governed embryogeny were erected, the most elaborate of which were presented by Crete (1963). Natesh and Rau (1984) reviewed and discredited all such schemes because careful investigators have always found enough variation to negate any formal embryogenic laws. Two examples described later in this chapter, apple (Meyer, 1958) and maize (Randolph, 1936), mention variable cell division patterns among the many proembryos they studied, but normal embryos nevertheless always resulted.

EMBRYOGENESIS IN DICOTS

The fast growing lettuce embryo (*Lactuca sativa*, Asteraceae), described carefully by H.A. Jones (1927), serves to illustrate the general features of dicot embryos as well as the relationship of the embryo to the common coenocytic/ multicellular type of endosperm.

Fertilization occurs about 3 hours after pollination, and by 6 hours the zygote is already flanked by the PEC coenocyte with a few nuclei. Shortly thereafter the zygote divides to form a typical small apical cell and large basal cell. The basal cell produces a small suspensor and also contributes some cells to the embryo proper. At about 9 hours post-pollination, both basal and apical cells divide, and by 10–12 hours the proembryo has 4 cells and is surrounded by the PEC coenocyte with numerous proliferating nuclei.

By 20 hours the proembryo has 8 cells, the suspensor 2, and both are embedded in the now multicellular endosperm phase. By 26 hours, barely more than a day after fertilization, there are already 12 proembryo cells and 3 suspensor cells, and more endosperm cells have formed. Jones did not mention numbers of proembryo cells for later stages, but by 34 hours the proembryo is in the early globe stage and has initiated the outermost epidermis-forming (dermatogen) layer. At 72 hours (3 days) there is a transition from the late globe proembryo to the early cotyledon embryo stage.

Jones showed later embryo stages without cellular detail and in relation to the whole developing seed. At 4 days post-pollination, immature heart stage cotyledons are seen and the multicellular endosperm is voluminous. The growth rate now slows, the suspensor degenerates and disappears, and by 7 days the embryo is at mid-cotyledonary stage, and only a small number of undigested endosperm cells remain. Considerable lettuce embryo expansion and endosperm absorption occurs during the next 4 days, and by day 11 the embryo is mature and all but a thin peripheral endosperm layer has been absorbed; the ultrastructure of these two persistent cell layers was described by R.L. Jones (1974).

The accounts of embryo development in field mustard (*Brassica campestris*) by Ahuja and Bhaduri (1956) and Rathore and Singh (1968) are less exhaustively detailed than the Jones account, but they show some differences from lettuce. Shortly after fertilization in the long, narrow embryo sac, the antipodal cells have already degenerated and the triploid PEC nucleus has divided once. Additional PEC nuclei proliferate while the zygote elongates and eventually divides to produce a tiny apical cell and an unusually elongate basal cell.

The next few cell divisions, like those of the lettuce proembryo, result in a tiny 4-celled proembryo sharply delimited from the bi-celled filamentous suspensor. The suspensor continues to grow faster than the embryo proper, and greatly exceeds the 3- celled proembryo in size. The terminal suspensor cell does not swell, unlike that of shepherd's purse (*Capsella*), the standard textbook example. The suspensor reaches maximum development at about this stage, and its proximal cell abutting the proembryo becomes the hypophysis (Greek, "growth under," referring to its position just below the proembryo), which will produce additional cells that contribute to both root cap and cortex of the radicle.

The proembryo at this stage initiates the dermatogen. The later globe proembryo stage, with most of the suspensor omitted; the outer dermatogen layer is evident and contributions of the hypophysis are labeled.

The elongate filamentous suspensor has caused the globe proembryo to encroach on, but not rupture, the enlarged coenocytic PEC with its numerous nuclei. The PEC in mustard begins its multicellular phase at a much later stage of embryo development than in lettuce. The late globe proembryo shows little internal differentiation, but the hypophysis has already contributed some cells that will become integrated with the developing radicle. Later growth involves cotyledon and hypocotyl development that resembles similar stages in lettuce and many other dicots. Endosperm has mostly been absorbed by the embryo by the time the seed is mature, and only a few peripheral cells remain around the curved embryo.

Before leaving herbaceous dicots, the "reproductive calendar" of Peruvian tomato (*Lycopersicum peruvianum*, Solanaceae) compiled by Pacini and Sarfatti (1978) should be mentioned. It provides a different way to correlate information about embryo and endosperm development, including times of filling and emptying of starch in the latter. Events in the surrounding ovary are also included, all listed in a convenient tabular form linked to days after pollination. The endosperm in Solanaceae is of the multicellular type (see *Nicotiana* example), so the PEC begins dividing directly into endosperm cells 3 days after pollination. It already has many cells by the time the zygote divides on the 5th day. The calendar indicates these and later events, up to maturity at day 21, in easily readable form. Embryo development requires about 16 days from fertilization to maturity.

Trees are, in general, more leisurely about seed development than the examples provided so far, but their stages of embryo development are similar. Apple (*Pyrus malus*) and pear (*Pyrus communis*) of the Rosaceae have been studied in detail and are good representatives of cultivated trees to take up next. The illustrations are from the pear, as presented by Osterwalder (1910). He also showed virtually identical stages for apple, and his work has been verified by Wanscher (1939) and Meyer (1958). Because the pear embryo requires at least two months to reach maturity, hour-byhour and day-by-day progress has not been described.

The deeply embedded crassinucellate pear embryo sac is of the common 7-celled *Polygonum* type, but the antipodal cells degenerate early (as in mustard described earlier), at or even before the pollen tube enters. One synergid is destroyed by the entering pollen tube and the second synergid then starts to degenerate. The coenocytic PEC already has a few nuclei while the zygote is still flanked by degenerating synergids and the pollen tube remnant. Nuclei continue to proliferate in the greatly elongating PEC while the proembryo becomes a short uniseriate filament of 5–8 cells.

Cells of the early filamentous proembryo of both apple and pear sometimes divide longitudinally to produce a biseriate proembryo, but other proembryos are less uniform. Variations in size and staining quality of the suspensor cells make some proembryos appear to have no recognizable suspensor; others obviously do, such as the swollen basal cells. Meyer (1958) examined 400–500 sectioned apple proembryos (a remarkable feat!) and verified Osterwalder's finding that early cell divisions are variable, as is the boundary between embryo proper and suspensor.

The globe stage of the apple proembryo typically originates from the uppermost four tiers of cells, although not always precisely so. These four tiers in the pear also usually initiate the globe stage, which is reached during the third week after fertilization. The elongate PEC coenocyte phase now has a large central vacuole and numerous nuclei along its periphery and extending across it just beyond the proembryo. Wanscher (1939) reported that the PEC nuclei proliferate in waves of synchronous division 2–3 days apart, each wave starting in the middle of the PEC and moving toward each end. At the stage shown in Figure 10.6G, 2000–3000 nuclei are present. After a final wave of PEC nuclear division, a short resting period follows, after which endosperm cell wall formation begins at the proembryo end of the PEC. Wan scher reported that the PEC never becomes completely subdivided into smaller cells; a portion always remains as a small coenocytic chalazal pocket. Osterwalder (1910) also showed this in an embryo sac at the young cotyledon stage.

Further development follows a common dicot pattern. The embryo grows at the expense of the now multicellular endosperm, with concomitant growth of the surrounding ovule and transformation of its integuments into the seed coat. The mature embryo in both apple and pear is still surrounded by several layers of endosperm cells. Perhaps the seemingly indistinct suspensor commented upon by Meyer (1958) and shown in some proembryos by Osterwalder (1910) indicates slower nutrient absorption, which in turn reflects the more leisurely growth rate of these embryos.

The four examples just described give an overview of dicot embryo development correlated with endosperm proliferation, with details concentrated on early embryo stages. As noted in these examples, dermatogen organizes itself early into an outer, epidermis-like layer, and cuticle may be secreted on its surface at a very early stage. Internal tissue differentiation, however, begins at the end of the globe stage, or sometimes just as cotyledons are initiated. The account by Miller and Wetmore (1945) for annual garden phlox (*Phlox drummondii*, Polemoniaceae) is readily understandable and representative of a great many dicots.

During the globe stage in phlox, just 5–6 days after fertilization, a central cylinder of somewhat darkly staining cells forms gradually in the hypocotyl of the embryo axis since there are no roots or cotyledons yet. At cotyledon initiation, a barely visible strand of procambium extends from the central column of the embryo axis into each of them. The root cap, initiated in the 7–8 day embryo at the basal end of the hypocotyl, is derived partly from the hypophysis, which also contributes cells to the meristem of the incipient root primordium beneath the root cap. The rather small suspensor now degenerates because the root cap has isolated it physically from the embryo proper. Further embryo development, to maturity at about 36 days, involves continued elongation of procambium within the embryo axis and cotyledons. Further meristematic activity in the embryo axis has formed a pith region distinct from procambium and a strand of elongate cells that will become protoxylem. There is patchy development of variously differentiated vascular tissue in mature cotyledons of different embryos.

Among dicot species in general, mature embryos exhibit variable vascular differentiation, both in extent and degree of differentiation, as revealed in some detail in the summary of embryo vasculature by Raghavan (1986). Because embryos in a seed are not subjected to any transpiration stress, this might help explain why vascular differentiation and interconnections are often made tardily.

Such patchy vascular development also occurs in young leaves and floral organs, which also develop in protected buds. The epicotyl or plumule (embryonic shoot above the cotyledonary node) also develops variably among species; the embryo in a dormant mature seed may lack leaf primordia beyond the cotyledons, or one to several immature leaves may have formed.

Embryo growth among dicots has been described in quantitative terms several times. It always follows a sigmoid curve when either length or volume is graphed against time. A representative example is jimson weed (*Datura stramonium*, Solanaceae), a weedy relative of tomato and potato. The embryo as a whole, as well as the hypocotyl and cotyledons separately, follows a sigmoid developmental curve. The zygote divides 1–2 days after pollination. The globe stage ends on the 11th day, when cotyledons are initiated. At 14 days, just at the start of the period of most rapid embryo growth, the ratio of water uptake to dry matter uptake has increased to its maximum; this rate declines thereafter as the embryo rapidly elongates to its mature length at 22 days. In other words, from 8–22 days the embryo becomes more and more dessicated, which is a necessary pre-condition for prolonged dormancy. Embryo volume at 14 days is barely measurable, but it increases along with length until the mature embryo at about 22 days has a volume of almost 2 mm^3.

EMBRYOGENESIS IN MONOCOTS

There are fewer cultivated monocot species than dicots, and monocot embryology has not been studied as extensively, except for cereal grasses, which have attracted attention because of their economic importance as well as certain embryonic structures regarded by many as unique. A consideration of monocot embryogenesis should not begin with a grass, however, but with a simpler and more representative example.

The coconut palm, *Cocos nucifera* (Arecaceae) seems on first thought to be unrepresentative because of its huge seed; nevertheless, it shows in rather uncomplicated fashion the main features of monocot embryo development. Coconut embryogeny was described in considerable detail by Haccius and Philip (1979), but because the embryo requires several months to grow to maturity, they did not correlate developmental stages with times after pollination or fertilization, nor did they include concomitant events in the endosperm.

The coconut zygote divides either transversely or obliquely, both planes producing a typical small apical cell and large basal cell. The synergid that was not penetrated by the pollen tube persists for some time into the early proembryo stage. As both apical and basal cells divide, the suspensor initially appears rather indistinct from the embryo proper; but at a later proembryo stage, the suspensor is distinct with larger and more vacuolate cells than those in the embryo proper. The suspensor also accumulates considerable starch. In the embryo proper, one of the cells derived from the original apical cell becomes more densely cytoplasmic than the rest, and it continues to divide while the other embryo cells do not. The progeny of this cell gradually replaces the rest of the cells, and eventually they form the entire globe stage.

The single cotyledon after initiation becomes a crescent-shaped ridge that extends around most of the flank of the globe proembryo. Cotyledon initiation marks the end of the proembryo stage and leaves the centrally located apical dome of the embryo free to initiate the shoot tip and leaf primordia (epicotyl or plumule). The growing cotyledon continues to extend its curved insertion laterally in both directions until it eventually completely encircles the embryo apex.

After the shoot apex of the embryo has enlarged to a certain size, the first foliage leaf primordium is initiated on its flank as a curved ridge that resembles initiation of the cotyledon. Both leaf and shoot apex are gradually pushed aside by an asymmetrical expansion of the cotyledon, one part of which bulges upward and occupies what appears to be a terminal position while the rest of the cotyledon continues to grow around and over the epicotyl. Six stages of cotyledon expansion during later embryo growth. The coconut radicle (root origin) arises from a flat, broad meristem that appears rather late in embryo growth at the pole opposite from the upper curvature of the cotyledon.

The shoot apex has by this stage been pushed aside by the expanding cotyledon from the usual position opposite the root pole that would be typical for a dicot. The seemingly strange appearance of the mature embryo therefore results largely from expansion of the cotyledon, which distorts the orientation of other embryonic structures. An account of embryo development in another important monocot, the African oil palm (*Elaeis guineensis*), by Fernandino et al. (1985), included less detailed anatomical information on cellular events and more on gross embryo growth and its duration, which complements the description of Haccius and Philip (1979) on coconut palm. The similar oil palm embryo takes about 80 days to reach maturity, at which time it consists of over 200,000 cells. The sigmoid growth curve of the oil palm embryo resembles the typical dicot growth curve except for its initial lengthy period of slow growth. A dramatic increase in oil palm embryo growth starts at about 44 days, when it has only 35–40 cells, which is during the same time that the PEC coenocyte begins forming internal walls to become multicellular endosperm. The timing of oil palm embryo development is similar to that of the coconut palm.

Monocots in general have abundant endosperm in the mature seed, most of which they do not utilize until seed germination. The cotyledon of most monocots is structurally adapted for pushing the rest of the embryo out of the seed while it remains partially or entirely embedded, absorbing and transferring digested and transformed endosperm to the seedling. In the coconut palm, with its enormous storehouse of endosperm (coconut milk and its more familiar final condition, coconut meat—the familiar shredded coconut on pastry), the curious humpbacked cotyledon elongates during germination and pushes the embryo axis out of the seed.

Most of the cotyledon remains within the seed, however, where it enlarges greatly as it absorbs endosperm. The seedling remains attached to the seed for a long time while it gradually uses what the cotyledon absorbs, digests, and translocates. Three examples of monocot embryos and their behavior upon germination will be described to emphasize that the monocot cotyledon is commonly a haustorium that only completes its development and function in the young seedling. Imagine that the humpbacked part of the coconut palm cotyledon just described is instead flared like the horn of a trumpet; this in miniature describes the cotyledon of the banana (*Musa* sp., Musaceae) as described by Gatin (1908).

The mature banana embryo looks like a tiny mushroom, the cotyledonary cap of which remains in the seed as a digesting organ, flattened against the endosperm like a suction cup, while the stalk part elongates and pushes out the embryo axis, which grows into the seedling. In the common onion (*Allium cepa*, Alliaceae), in contrast, the cotyledon has the form of a slender flexible cylinder, which elongates and accommodates itself to the seed by curling up within the endosperm as it digests and softens it. The proximal part of this cotyledon elongates upon germination, thereby pushing out the embryo axis, which becomes the seedling. As in banana and coconut, the onion seedling is supplied for a certain period by nutrients absorbed from the endosperm by the haustorial distal part of its cotyledon.

Even before germination, the slender, curved shape of the embryo suggests the later curved cotyledon. Among grasses the cotyledon is called the scutellum (Latin: "little shield"); it also serves to digest and translocate endosperm nutrients to the seedling (Negbi, 1984). Although functionally just like the cotyledon of the monocots described previously, grasses differ because the entire cotyledon (scutellum) remains in the seed. Elongation occurs instead in the so-called mesocotyl (the internode between scutellar node and coleoptile) of the embryo axis and also in its protective conical coleoptile, which is the first true leaf.

The scutellum of some grasses increases its surface area while embedded in the endosperm by forming tubular extensions from epidermal cells that appear similar to root hairs. The surface of the maize scutellum, however, merely becomes convoluted, with deep clefts that are possible sites of enzyme secretion. In oats (*Avena*) and its relatives the scutellum itself elongates along with forming epidermal extensions, thereby expanding within the seed to absorb endosperm rapidly and efficiently. The grass scutellum is therefore a haustorial cotyledon like those of other monocots.

Among monocots, grass embryos have received by far the most attention, particularly those of cereal grasses. Maize (*Zea mays*) is described here as a representative example. Among numerous investigations of maize the classic study by Randolph (1936) is comprehensive and provides most of the information and illustrations for the following description. Abbe and Stein (1954) and more recent contributions by Van Lammeren (1986, 1987) are valuable supplements to Randolph's account. The mature maize embryo sac just before fertilization can be seen in previous chapter. The zygote divides 10–12 hours after fertilization. Cell walls resulting from early divisions in the proembryo may be variously oriented, and Randolph illustrated several of them. The suspensor merges broadly with the proembryo, continuing to enlarge along with it in volume but not in cell number. During days 5–6, the proembryo elongates and becomes club-shaped.

During day 7, the proembryo proper develops a recognizable protoderm, which extends to the suspensor boundary. The suspensor has also added cells and increased its volume. The first sign of internal differentiation occurs about 10 days after pollination, in the form of a group of densely protoplasmic and actively dividing cells at the front of the embryo, slightly below the tip, which becomes the apical meristem destined to produce the coleoptile, foliage leaves, and reproductive branches. Note that the scutellum (cotyledon) appears to be terminal.

By day 14 there are two meristems, one associated with the shoot tip, the other with the root primordium, and the suspensor is at maximum size. The coleoptile (Greek for "featherlike sheath") has now been initiated; it can be considered simply as the first leaf formed after the cotyledon (scutellum), although it differs from other leaves because it has the form of an elongate tube closed at the apex except for a slit. Further growth of the maize embryo by day 20 involves enlargement of all parts (except the suspensor) and initiation of the first leaf primordium after the coleoptile. Just as the apical meristem becomes covered by the protective coleoptile, so the root meristem is sheathed by another unusual structure, the coleorhiza (Greek for "root sheath"). This structure occurs in many, perhaps most, grasses.

During the following 21 days or so, all parts of the embryo increase in size, and all of the five to eight leaves (depending on variety) that will be present in the mature plant are initiated. Considering factors such as varietal genotype and growing conditions, the maize embryo takes about six weeks (42–45 days) to develop from the zygote to its final form within the dormant kernel.

Abbe and Stein (1954) verified much of Randolph's information and complemented it with unique illustrations of embryo development, which depict each stage both in typical median longitudinal sectional view and in 3-dimensional face view. One can see that the scutellum does indeed shield the embryo axis by wrapping itself around it like an overcoat (stages 4–6). And the sheathing coleoptile can be seen to gradually envelop the epicotyl and all of its parts (stages 1–6), leaving only a tiny pore through which the epicotyl will emerge upon germination. Van Lammeren (1986) showed similar views of maize embryo growth by scanning electron microscopy. When the grass embryo is seen in three dimensions, one can appreciate the versatility of the monocot cotyledon compared to the more leaflike dicot cotyledon.

A study that included both physiological and morphological changes was conducted by Johri and Maheshwari (1966) on embryo development of zephyr lily (*Zephyranthes lancasteri*, Amaryllidaceae), a cultivated monocot related to amaryllis. They chose to omit a detailed description of cellular development and instead showed the developing embryo in relation to the growth of the entire ovule and seed in a series of stages, at several time intervals after pollination, until seed maturity at 18 days. Note that the coenocytic PEC initiates the multicellular endosperm phase on day 7, is mostly cellular by day 9, and is completely cellular by day 11. As is typical of monocots, the mature embryo with its elongate terminal cotyledon is embedded in a copious amount of endosperm, which it utilizes during germination.

Johri and Maheshwari (1966) also presente‹ graphically some related physiological measures for zephyr lily. During development the ovule increases in length about 10 times and the embryo about 600 times, and the embryo follows a typical sigmoid developmental curve. The fresh weight of an ovule increases until just about when the cotyledon begins to elongate, and then fresh weight declines rapidly. This graph also shows a more steady, modest increase in dry weight all through development, except during the last few days. Water content is initially almost steady, but it increases with the early increase in fresh weight, begins to decline as endosperm becomes multicellular, and declines more rapidly as the cotyledon elongates.

This follows approximately the pattern established for *Phaseolus*, as discussed in Raghavan (1986), thus indicating that these processes leading to dormancy are probably similar in dicots and monocots. Respiration in the zephyr lily ovule—as measured by oxygen uptake—peaks twice, the first coinciding with the start of PEC subdivision into cells, and the second and larger peak at the time of cotyledon elongation. The respiration peaks indicate maximum expenditure of energy at these times.

A similar sigmoid curve for embryo development was described graphically for onion (DolezØel et al., 1980). Their graph begins on the 9th day with the globe stage proembryo and the start of multicellular endosperm; before day 9, nuclei of the coenocytic PEC have divided slowly until the zygote divides on about the 5th day. The embryo grows most rapidly from day 18 to day 30. Maturity is reached at about 36 days after pollination.

NUTRITION OF THE EMBRYO

The embryo in the ovule acquires its nutrition in a heterotrophic (non-photosynthetic) manner, which means that it absorbs energy-rich substances such as sugars through the epidermis of some or all of its developing body. At early stages the suspensor (if it is a functional one) seems to perform much if not all of this function. This remarkable ability is glossed over in the literature by statements that the embryo "absorbs" endosperm during development and/or during germination. Only shortly after germination does the seedling become autotrophic—i.e., turn green and photosynthetic (but see possible exceptions in the later section on "green embryos").

A consideration of events during germination is beyond the scope of this book. Heterotrophic nutrient absorption by the young seedling has attracted virtually all of the attention, and earlier in this chapter the monocot cotyledon was shown to be adapted to serve this purpose. Details of enzyme secretion and its action on endosperm digestion have been studied widely and intensively. Except for the conspicuous, even spectacular, anatomical transformations of the body and epidermis of the scutellum of certain grasses, and the obviously haustorial cotyledon of other monocots such as palms, there has been little correlation of anatomical and ultrastructural information with that from physiology.

Before the seed germinates, little attention has been paid to how the developing embryo absorbs what it needs from the endosperm in order to grow from a zygote to its mature dormant state. At late embryo stages there is indirect evidence implicating the cotyledon(s) from reports of polyploidy in cotyledon cells: 4n–8n is common, levels of 12n have been reported in cotton, 16n in maize scutellum, and 32n in cotyledonary epidermal cells of the monocot *Gibbaeum* (Nagl, 1978). Such heightened levels suggest intense metabolic activity by cotyledon epidermal cells, and it is reasonable to speculate that they are involved in enzyme secretion and related processes that convert endosperm to simpler molecules for transfer to the cotyledons.

Smart and O'Brien (1983) studied structural changes in wheat endosperm as the embryo develops. They noticed that a densely cytoplasmic sheath of "modified" endosperm cells forms around the early globe stage proembryo. These modified cells disappeared gradually, leaving the older but still immature embryo separated from the nucellus by what appeared to be an empty space; in addition, several layers of nearby nucellar cells had intact walls but appeared devoid of cell contents. Smart and O'Brien interpreted these changes as evidence that the young wheat embryo is a "powerful sink" that draws off nutrients from these nearby tissues.

Cell wall ingrowths that greatly expand cell membrane area were seen at a later stage in nucellar cells, followed by starch accumulation in the mid-region of the embryo and later in the coleoptile apex and the coleorhiza. These structural changes provide indirect evidence of how the wheat embryo possibly drains nearby cells to obtain nutrients for its own growth. It is of interest that Smart and O'Brien found no special relationship between the suspensor and adjacent tissue, which one would have expected; instead, suspensor cells were similar in size and shape to those of the embryo proper, although they were somewhat more vacuolate.

A zone of "degraded" endosperm cells has also been shown around the mature embryo of celery (*Apium graveolans*, Apiaceae) by Jacobsen and Pressman (1979). These are endosperm cells from which the cytoplasm has been absorbed in some manner by the developing embryo, and these cells appear strikingly different from the remaining mass of endosperm that is digested later during seed germination. They reported that a gibberellin probably stimulates endosperm digestion during germination, but they did not speculate about how the earlier degraded endosperm cells were digested.

The question of when the cuticle forms on an embryo is pertinent to a consideration of when and how the embryo absorbs nutrients, especially during later stages after the suspensor becomes nonfunctional. Bruck and Walker (1985a,b) showed a cuticle on globe-stage proembryos of rough lemon, *Citrus jambhiri*, and Lackie and Yeung (1996) detected a cuticle in the carrot proembryo soon after the protoderm formed at the globe stage. But in both investigations the suspensor was reported to lack a cuticle. A cuticle on the mature embryo has been reported from alfalfa, *Medicago sativa* (Singh, 1977) and the cashew nut tree (*Anacardium occidentale*, Anacardiaceae) (Wood and Vaughan, 1976). In the two latter investigations the embryo cuticle was reported to be very thin compared to the cuticle found on aerial parts of mature plants, which is understandable since a thick cuticle would be a formidable barrier to extracellular heterotrophic nutrition.

Another clue to nutrient absorption was provided by Paramonova (1975), who reported that mature pea (*Pisum*) cotyledons have a thin cuticle, and that pores up to 1.5 ¼m in diameter extend most of the way through the outer wall of epidermal cells. In a later paper Paramanova (1981) showed that numerous cytoplasmic channels extend into the cell wall from the inte rior of epidermal cells but they are gradually closed off late in cotyledon development. He regarded these channels as pathways through which nutrients move from the endosperm into the cotyledons.

INDUCTION OF DORMANCY

The embryo has no intrinsic urge to go dormant. It simply continues growing into a seedling unless dormancy is imposed on it. There are numerous reports of seeds germinating and growing out of the fruit while still on the plant, often in response to very humid conditions. One report described a citrus seedling 10 cm long, with fully green chloroplasts, growing entirely within a lemon fruit (Whatley and Price, 1983). There are also mutations known among crop plants that eliminate dormancy and allow the embryo to continue growing. Mangroves include several species of dicot trees inhabiting brackish or salt water that are well known for naturally allowing seeds to germinate and grow into seedlings while still attached to the parent tree, after which they drop taproot first, like spears, into the tidal mud flat below and establish themselves rapidly. All of these kinds of nondormancy are examples of vivipary ("live-bearing"), the same term used for animals that bear their young live instead of laying eggs. Dormancy is the common termination to embryo development within a seed, however, and its causes are several.

The easiest kind of dormancy to understand is called coat enhanced, which is imposed by either a hard seed coat, by a resistant layer of endosperm, or by the two working together. Such physical restraint of the embryo also prevents oxygen and water from diffusing into the seed. Breaching this physical barrier is all that is needed to let the embryo resume growing. True dormancy is considered to be suspended embryo growth resulting from an inhibitor produced by the embryo, or imposed on it, which can be overcome only by special conditions (such as a cold treatment) beyond simply putting it into a favorable germination milieu.

The imposition of developmental arrest and true dormancy, and the removal of these restraints during germination, are topics of great practical importance as well as intrinsic interest. Several books on seed physiology and germination deal with these matters in great detail—for example, Bewley and Black (1978) and Kigel and Galili (1995)—and Bewley 202 *Flowering Plant Embryology* (1997) is a concise review of dormancy and germination.

With respect to explaining how a developing embryo becomes dormant, a detailed investigation by Walbot (1978) that combined embryonic stages with physiological measures in the bean (*Phaseolus vulgaris*) deserves mention. Following fertilization, the bean proembryo reaches the globe stage in only 2 days, and by day 7 the cotyledon primordia appear. From then until day 16, the cotyledons grow mainly by adding new cells, their number doubling about every 24 hours. The rate of cell division then slows, and it takes 8 days until the cell number is doubled again by day 24. The final doubling of cell number occurs during days 24–32. The embryo is mature and dormant by day 36. After cotyledons are initiated on day 7, embryo dry weight doubles about every 4 days, thus wall material accumulates in addition to new cells. Water content per embryo cell increases until day 28, but deposition of dry matter renders the water a smaller component of total embryo weight from day 18–28. After day 28 there is an actual water loss until maturity at day 36.

During early development up to about day 18 the bean embryo grows rapidly, and levels of growth promoter substances remain high. From day 18–28, conditions actually favor precocious germination of the embryo, but also during this time period the concentration of abscisic acid increases, and this inhibitor of RNA synthesis also inhibits growth of the embryo. Walbot (1978) felt that the beginning of water stress late in embryo maturation stimulates abscisic acid production; however, after day 28 the seed coat and the peripheral endosperm layer both harden. These last two events, combined with progressive dessication of the seed, reduces the level of abscisic acid. Although abscisic acid is the main inhibitor of the embryo at stages when water levels are still conducive to growth, dessication appears to be the final factor that prevents further growth.

A later study on a viviparous mutant of maize by Neill et al. (1987) provides support for Walbot's hypothesis. Viviparous maize embryos did not become dessicated late in seed development, and total abscisic acid levels were low; therefore, the grains germinated on the plant. By adding abscisic acid they were able to prevent precocious germination. The *Phaseolus* and maize examples just described illustrate the interplay between the embryo and its surrounding tissues that occurs during embryo maturation and the induction of dormancy. Variations exist among different species, of course, and controversies about what causes dormancy have developed in the literature. At least some problems of interpretation have resulted because many investigations have been conducted using excised embryos grown in culture rather than in the intact seed. Raghavan (1986) has discussed these matters, as has Bewley (1997), but further discussion of these matters is beyond the scope of this book.

GREEN (CHLOROPHYLLOUS) EMBRYOS

One would not expect a dormant embryo in the seed to be green or capable of photosynthesis. Green embryos are surprisingly widely distributed, however, and most people have seen the green embryos in mature seeds of peas, beans, and citrus fruits. Published reports and descriptions of green embryos among angiosperms were collated by Dahlgren (1980) from more than 1,000 species. He found that green embryos generally occur in species with little or no endosperm in the mature seed, and species with endosperm-rich mature seeds usually have nongreen embryos.

For example, legumes that store endosperm mostly in their cotyledons tend to have green embryos (e.g., peas, beans), whereas the endosperm-rich grasses have white embryos. But exceptions occur. Members of the Rosaceae, for example, have non-green embryos whether endosperm is present or not in mature seeds, and embryos of sunflower and their relatives

(Asteraceae) are non-green although the mature seeds lack endosperm.

Are green embryos capable of photosynthesis, and thus possible contributors to their own development? Whatley and Price (1983) reported that chloroplasts form early in embryos of some species, and then dedifferentiate to proplastids (rudimentary chloroplasts lacking chlorophyll), followed by redifferentiation to functional chloroplasts at seed maturity. Chloroplasts in other species differentiate at the immature cotyledon (heart) stage, and the embryo remains green thereafter. In still other species the chloroplasts do not differentiate until the seed begins to germinate. Whatley and Price (1983) determined that light in the red wavelength, which is necessary to convert protochlorophyll to chlorophyll, is capable of reaching both seeds and the embryos within them in lemon fruits.

They also found that light can penetrate, at least feebly, into the interior of many other fruits and seeds. It is possible that chloroplast differentiation does little or nothing for developing embryos, and merely pre-adapts them for rapid growth when exposed to light upon germination. The lack of chlorophyll development in most embryos in endosperm-rich mature seeds probably indicates that endosperm remains the source of nutrition for a relatively long period of time after germination.

POLYEMBRYONY

One or more asexually produced embryos can occur in addition to, or in place of, normal sexual embryos. This phenomenon, called polyembryony or somatic embryogenesis, is known from about 250 species in 59 families of both dicots and monocots.

The extra embryos form from cells that can arise spontaneously from any part of the ovule, even budding off from the surface of the normal sexual embryo itself. Nucellus, inner integument, synergid cells, antipodal cells, endosperm, and suspensor have all been reported to generate such embryos.

Tisserat et al. (1979) provide the most comprehensive review. The asexual embryos may develop to maturity along with the sexual embryo, or the sexual embryo may die off and be replaced by one or more asexual embryos. *Citrus* is the best known genus for polyembryony, but eight other genera of Rutaceae also exhibit it; all extra embryos in this family arise from the nucellus.

Mango (*Mangifera*, dicot family Anacardiaceae), onion, and tobacco are a few other cultivated plants in which polyembryony is known. Knowledge of the causes of polyembryony is only speculative, as admitted by Lakshmanan and Ambegaokar (1984), but it has been induced through cytological abnormalities and application of certain chemicals. Such induced asexual embryos have been used in plant breeding to obtain genetically uniform plants of various ploidy levels. The adaptive significance of polyembryony for plants occurring in nature is, however, obscure.

APOMIXIS

Apomixis (Greek, "without mixing") is a form of asexual reproduction that mimics sexual reproduction by using some or all of the sexual apparatus, except fertilization, to produce an asexual embryo and seed. This distinguishes apomixis from polyembryony, in which proliferating cells arise from tissues in and around the embryo sac and take the form of one or more embryos that accompany or replace the primary sexual embryo.

Apomixis can produce viable seeds because the embryo sac (female gametophyte) is produced without meiosis; thus the egg cell already has both male and female genomes, as does the PEC nucleus. This form of asexual reproduction is widely distributed among both dicots and monocots, most commonly occurring as an alternate to normal sexual reproduction under certain conditions.

Its causes are several—for example, high ploidy levels (diploids are rarely apomictic) and weather changes. Apomixis is often said to have an advantage in weedy species (dandelion is probably the best known apomict), which can spread into new habitats without requiring nearby plants for pollination and fertilization.

Versions of apomixis range from no pollination required, to the need for pollen or pollen tubes in the style, to the actual entry of sperm into the egg cell (but no merger of nuclei) to stimulate it to divide and form an embryo; many intermediate possibilities are also known. Some forms of apomixis even require merger of the second sperm to stimulate the PEC to form endosperm. Gustafsson (1946), Battaglia (1963), and Nogler (1984) collectively provide an extensive and exhaustive nomenclature to cover the many variations.

As a plant breeding strategy, apomixis can be advantageous because it produces genetically identical seeds. Unfortunately, apomictic plants are difficult to induce reliably or even to identify, and commercial use is not yet at hand. Bashaw (1980) discussed the advantages and disadvantages of breeding with apomicts, and more recent reviews do not report that the problems have been solved. Recent analyses and detailed discussions of genetic and molecular aspects of apomixis are by Mogie (1992) and Koltunow et al. (2002).

SUMMARY: EMBRYO AND SEED

The embryo is the new sporophyte, and it establishes the tissue and organ architecture that will prevail during germination and subsequent vegetative growth of the plant. The suspensor is a unique but ephemeral embryo structure with a great range of size, shape and DNA levels, indicating that its level of physiological involvement in proembryo development probably also varies considerably among species. The cotyledons of dicots may absorb some or all of the endosperm reserves before dormancy, whereas the single cotyledon of most monocot embryos is adapted instead to be haustorial, absorbing most of the endosperm during and after germination for the benefit of the seedling. Extra embryos, or even a replacement embryo, can be produced asexually in many species by polyembryony or apomixis.

A mature embryo will continue to grow without interruption into a seedling unless dormancy is imposed on it to allow the seed to complete its development. This inherent property of the embryo means that it is the young sporophyte and not something set entirely apart from it, despite the occurrence of a couple of exceptional structures (cotyledon, suspensor) and a heterotrophic mode of nutrition.

The seed coat, fruit development, and fruit and seed dispersal are each complex topics with endless variations adapted to habitat and other ecological and physiological requirements for existence and continuation of the life cycle of individual species. These topics are, however, beyond the scope of this book. A good entry into these aspects, especially as related to cultivated plants, is Kigel and Galili (1995).

11

Biochemical and Molecular Genetic Aspects of Floral Scents

INTRODUCTION

The chemical composition of floral scents has been extensively investigated for hundreds of years because of the commercial value of floral volatiles in perfumery. More recently, several ecological studies have examined the roles of floral scent in the biology of the plant. However, in contrast to the chemical emphasis of the perfumers and the organismal emphasis of the ecologists, until recently, there have been few studies concerning the biochemical synthesis of floral scent compounds and the enzymes and genes that control these processes. In fact, our recent investigation into the biogenesis of floral scent production in Clarkia breweri, an annual plant native to California that served as our model organism, and our even more recent work on the scent of the cultivated snapdragon (Antirrhinum majus) represent the only examples to date (to our knowledge) in which isolation of enzymes and genes involved in the de novo synthesis of scent compounds in the flower have been reported. In this Update we review the research leading to our work, report our findings, and discuss implications for future directions of the field.

FLORAL SCENTS ARE IMPORTANT FOR PLANT FITNESS

Many plants emit floral scents, and such scents can attract a variety of animal pollinators, mostly insects. When present, scent is often the dominant means of long-distance attraction, particularly in moth-pollinated flowers, which are searched out and visited at night. Floral fragrances vary widely among species in terms of the number, identity, and relative amounts of constituent volatile compounds.

Although little is known about how insects respond to individual components found in floral scents, it is clear that insects are able to distinguish between complex floral scent mixtures, and that discriminatory visitation based on floral scent has important implications for plant reproductive success (Pellmyr, 1986). Since floral scent can be crucial in ensuring fertilization, and therefore in determining seed or fruit set, the presence or absence of a scent attractive to the locally available insect pollinators may have a substantial impact on the yield of agronomically important crops (Free, 1970).

Plants imported into a new environment by humans may be especially disadvantaged in this regard, as they have not co-evolved with the local pollinators (Traub et al., 1942). Even if a local pollinator is attracted to the flowers, it may not be physically suitable to be an effective pollinator. On the other hand, pollinators that may have the appropriate physique (by chance) may not be successfully attracted to the plant (Herrera, 1987).

Plants did not naturally evolve to produce their scent for the benefit of humans; nevertheless, it is clear that humans find an aesthetic value in certain types of floral scents, and the presence of floral scent may have contributed to the decision by humans to cultivate and propagate specific plant species. While there is certainly a wide variation in human taste, most people prefer the scents of bee-pollinated and, especially, moth-pollinated flowers, which they often describe as "sweet-smelling" (Knudsen and Tollsten, 1993). Unfortunately, very few plants are currently cultivated primarily for their scent. Moreover, a large number of commercial flower varieties have lost their scent during the selection and breeding processes due to, on the one hand, a focus on maximizing post-harvest shelf-life, shipping characteristics, and visual aesthetic values (i.e. color, shape), and on the other hand, to the lack of selection for the scent trait.

Some volatile compounds found in floral scent have important functions in vegetative processes as well. They may function as attractants for the natural predators of herbivores (Pare and Tumlinson, 1997) or as airborne signals that activate disease resistance via the expression of defense-related genes in neighboring plants and in the healthy tissues of infected plants (Shulaev et al., 1997). They may also serve as repellents against herbivores (Gershenzon and Croteau, 1991).

DETERMINATION OF FLORAL SCENT COMPOSITION

Until recently, investigations concerning floral scent have concentrated mainly on determining the chemical composition of floral fragrances. For this purpose, the "headspace" collection method was developed. In this procedure, a flower that is still connected to the rest of the plant is placed inside a glass chamber and its emitted volatiles are collected by continually purging the air inside the chamber through a polymer mesh that binds these volatiles. After a fixed period of time, the volatiles bound to the polymer are extracted with an organic solvent (a variation of this procedure, the highly sensitive solid phase microextraction method, allows for "instant" sampling of headspace volatiles; Matich et al., 1996). The solution is then injected into a gas chromatograph, which separates the different volatiles, and each volatile is identified by mass spectrometry. These investigations have determined that floral scents are almost always a complex mixture of small (approximately 100-250 D) volatile molecules and are dominated by monoterpenoid and sesquiterpenoid, phenylpropanoid, and benzenoid compounds. Fatty acid derivatives and a range of other chemicals, especially those containing nitrogen or sulfur, are also sometimes present (for review, see Knudsen et al., 1993).

LOCATION OF EMISSION OF FLORAL SCENT COMPOUNDS

The question of from which part of the flower the various scent components are emitted is a technically difficult one to answer. The identification of specific compounds inside floral organs is not sufficient to prove that these compounds are in fact emitted directly from such organs. The identification of anatomical features designated as "scent glands" by staining with stains that have affinity to aliphatic compounds (Curry, 1987) is also problematic, because this methodology again does not involve the actual measurement of scent emission from such anatomical features (nor from any other part of the flower). The best approach has been to conduct headspace analysis on flowers from which certain parts have been excised.

These investigations have found that while the same floral scent components are often emitted from several parts of the flower (although not necessarily at the same amount or rate), specific compounds may sometimes be emitted from only a subset of the floral organs involved in total scent emission (Dobson et al., 1990; Pichersky et al., 1994). However, these results should also be viewed with caution, because the experiment involves injury to the flower (the removal of organs) and may therefore lead to changes in emission profiles. The use of appropriate controls and comparisons with whole-flower emission profiles should be employed to identify such changes should they occur (Dudareva et al., 1998b).

TEMPORAL AND PHYSIOLOGICAL VARIATIONS IN FLORAL SCENT EMISSION

There are many reasons why plants need to, and often do, vary the floral scent they emit during the lifespan of the flower, both in total output and in specific composition. It is to the advantage of the plant to have its scent output at maximal levels only when its potential pollinator is active.

Thus, flowers that are pollinated by nocturnal insects such as moths tend to have maximal scent output in early evening (Loughrin et al., 1990), although this is not always the case (Pichersky et al., 1994). Aside from the obvious consideration of conservation of energy, it is also to the advantage of the plant not to attract more general pollinators, who might disperse its pollen non-productively. However, a certain amount of visitation by generalist pollinators might have some benefit to the plant as "insurance," in case its specific pollinator is rare or absent.

It has been suggested that specific compounds in floral bouquets are designed to attract specific pollinators, and that some flowers can change their floral scent composition over time to attract more general pollinators if the flower has not yet been pollinated. In addition, it has been hypothesized that some flowers emit certain chemicals designed to repel insects that are non-beneficial to the plant (e.g. the so-called pollen or nectar "thieves," or generally destructive insects). Although specific changes in floral scents that follow diurnal, nocturnal, or circadian rhythms (and separate patterns may apply to different compounds in the same flower) or some other specific program over the lifespan of the flower have been well documented (Dudareva et al., 1999), our knowledge of how insect pollinators respond to specific floral volatiles is so rudimentary that it is not yet possible to assign specific adaptive values to such changes.

It should also be remembered that insects are capable of associative learning, and that the interests of the insect pollinators and those of the plants are not completely complementary, so there is a certain amount of "cheating" going on in these relationships. Sometimes the plants have the upper hand, as in the case of the Ophrys genus, where the insects are lured into visiting and pollinating the plants with the ruse of pheromone mimicry by floral scent (as well as morphological mimicry) (Borg-Karlson and Tengo, 1986), but often the plants are on the losing end.

Nonetheless, it often benefits the flower to advertise truthfully when true rewards such as nectar or pollen are available (and their availability should coincide with the time of maturity of the male and female parts), and to stop advertising when the rewards are not available. This is because the insects will learn to associate the scent emanating from the flower with the rewards they get when they reach the flower, and will therefore continue to seek out and visit flowers with the same scent. On the other hand, even if the insect has been lured once by the scent to come and pollinate the flower when no rewards are available, it is less likely to come again to another flower of this kind.

In addition, once a flower has been sufficiently pollinated, post-pollination changes in scent emission could prevent additional, possibly destructive, visits to the pollinated flower, and could therefore also increase the chances of visitation to unpollinated flowers. Indeed, in most cases analyzed, the scent of flowers is markedly reduced soon after pollination, and in some cases it has been found to change qualitatively. The cessation of scent emission is often, but not always, accomplished simply because the petals (which usually constitute the bulk of the flower and the main source of scent emission), stigma, and style senesce and wilt.

Finally, the effect of temperature on fragrance emission has generally received even less attention, but it has been shown that temperature has a strong effect on the quantity of fragrance. For example, total emission of fragrance from Trifolium repens L. flowers was 58% higher at 20°C than at 10°C and all compounds of floral scent were affected by the change in temperature (Jakobsen and Olsen, 1994). It is not clear if the increase in emission is due solely to the greater volatility of these compounds at the higher temperature, or if it is also due to biological processes, including increased synthesis.

BIOSYNTHESIS OF FLORAL SCENT COMPOUNDS

As mentioned in the introduction, there have been few studies concerning the biochemical synthesis of floral scents, and our recent investigation into the biogenesis of floral scent production in our model organisms C. breweri and snapdragon represent, to our knowledge, the only examples to date in which isolation of enzymes and genes responsible for the formation of scent volatiles in the flower have been accomplished. These results are discussed below. Fortunately, too, many of the volatiles found in floral scents are also synthesized in vegetative tissues under specific conditions (mostly for defense purposes), and some information concerning their biosynthesis is also available. While it cannot be taken for granted that the synthesis of such compounds in vegetative tissue will in all cases be identical (i.e. same reactions, same enzymes) to their synthesis in flowers, and while there is currently no evidence that volatiles synthesized in vegetative tissues are transported into flowers, it is nevertheless instructive to review this information.

Terpenes, especially monoterpenes such as linalool, limonene, myrcene, and trans- -ocimene, but also some sesquiterpenes such as farnesene, nerolidol, and caryophyllene, are common constituents of floral scent. They are also often found in vegetative tissues, where they serve mostly as defense compounds. In work done mostly with vegetative tissue, but also with daffodil petals, it was found that monoterpenes are synthesized in the plastidic compartment. In this cellular compartment, isopentenyl pyrophosphate (IPP) is derived from the mevalonate-independent "Rohmer" pathway (Lichtenthaler et al., 1997). IPP can be isomerized to dimethylallyl diphosphate (DMAPP), and one molecule of IPP is condensed with one molecule of DMAPP in a reaction catalyzed by the enzyme geranyl pyrophosphate synthase (GPPS) to form GPP, the universal precursor of all the monoterpenes. Similar work with vegetative tissue has revealed that in the cytosol, IPP is derived from the mevalonic acid pathway (McCaskill and Croteau, 1998), and two molecules of IPP and one molecule of DMAPP are condensed in a reaction catalyzed by the enzyme farnesyl pyrophosphate synthase (FPPS) to form FPP, the universal precursor of all the sesquiterpenes (McGarvey and Croteau, 1995).

In the last few years, genes encoding the enzymes responsible for the synthesis of many monoterpenes and sesquiterpenes have been identified and characterized (Bohlmann et al., 1998). However, to date, only the enzyme that catalyzes the formation of the acyclic monoterpene linalool has been characterized in floral tissue. C. breweri flowers emit copious amounts of S-linalool from the petals, stigma, and style (the stigma and style also emit large amounts of linalool oxides), and we were able to demonstrate that linalool was synthesized from GPP in a one-step reaction catalyzed by a monomeric enzyme linalool synthase (LIS) (Pichersky et al., 1994). We were also able to purify LIS from C. breweri stigmata by employing several chromatographic techniques (Pichersky et al., 1995) and to obtain peptide sequences that allowed us to isolate a LIS cDNA clone from a C. breweri flower cDNA library (Dudareva et al., 1996).

The phenylpropanoids, which are derived from Phe, constitute a large class of secondary metabolites in plants. Many are intermediates in the synthesis of structural cell components (e.g. lignin), pigments (e.g. anthocyanins), and defense compounds. These are not usually volatile. However, several phenylpropanoids whose carboxyl group at C9 is reduced (to either the aldehyde, alcohol, or alkane/alkene) and/or which contain alkyl additions to the hydroxyl groups of the benzyl ring or to the carboxyl group (i.e. ethers and esters) are volatiles. Our work with C. breweri flowers has now resulted in the identification and characterization of three enzymes that catalyze the formation of floral volatiles from this group: (iso)methyleugenol, benzylacetate, and methylsalicylate. The enzymes are, respectively, S-adenosyl-L-Met:(iso) eugenol O-methyltransferase (IEMT), acetyl-CoA:benzylalcohol acethyltransferase (BEAT), and S-adenosyl-L-Met:salicylic acid carboxyl methyltransferase (SAMT) (Wang et al., 1997; Dudareva et al., 1998a, 1998b; Wang and Pichersky, 1998; Ross et al., 1999). In addition, we have identified and characterized the enzyme S-adenosyl-L-Met:benzoic acid carboxyl methyltransferase (BAMT), which catalyzes the formation of methylbenzoate in snapdragon flowers (Bushue et al., 1999). cDNAs encoding all of these enzymes have also been characterized.

REGULATION OF SCENT BIOSYNTHESIS

In both C. breweri and snapdragon flowers, emission of the bulk of the volatiles occurs from the petals. Identification of enzymes responsible for the formation of these volatile compounds allowed us to determine how the levels of enzymatic activities are distributed in different floral parts and how they vary during flower development. When activity levels are calculated per total weight of each organ, the highest levels of activity of all these enzymes are found in the petals (Dudareva et al., 1999). Other parts of the flower, however, also contain detectable levels of activity, and the stigma actually contains higher levels of LIS specific activity (but because the mass of the stigma of C. breweri is so small compared with the mass of the petals, LIS in the petal still comprises the majority of activity present in the flower).

The specific types of cells expressing the genes encoding LIS and IEMT were determined by in situ hybridization. The results indicate that in C. breweri flowers, these scent genes are expressed uniformly and almost exclusively in cells of the epidermal layer of petals and other floral parts (Dudareva et al., 1996). Volatile compounds produced in epidermal cells can apparently escape directly into the atmosphere after being synthesized.

C. breweri flowers, despite being moth-pollinated, do not show marked differences in emission between day and night. Snapdragon flowers, on the other hand, are bee-pollinated and have a marked peak of emission during the day. Both types of flowers follow a long-term pattern in which emission peaks within a few days of anthesis and then declines gradually. In C. breweri, the activities of scent enzymes follow two different patterns. The activities of the first group of enzymes, represented by LIS and SAMT, increase in maturing buds and young flowers, peaking about 12 to 24 h ahead of peak volatile emission. LIS and SAMT activities then decline in old (5-d) C. breweri flowers, but remain relatively high (40%-50% from the maximum level) even though emission of linalool and methylsalicylate has practically ceased.

The activities of the second group of enzymes, represented by IEMT and BEAT, show little or no decline at the end of the lifespan of the flower, although, again, emission of methyleugenol, isomethyleugenol, and benzylacetate virtually cease. A minor difference in developmental profiles of the latter two enzymes is that IEMT levels peak on d 1 of anthesis and stay stable afterward (Wang et al., 1997), whereas BEAT activity does not peak until the 4th d after anthesis (Dudareva et al., 1998a). The BAMT enzyme from snapdragon flowers appears to belong to the first group, since its activity declines at the end of the lifespan of the flower (9-12 d after anthesis, snapdragon flowers are longer-lived than those of C. breweri).

The causes and consequences of appreciable levels of activity of biosynthetic enzymes in old flowers, without concomitant emission of the volatile products, are unknown. Although it is possible that the biosynthetic pathways in which these enzymes participate are blocked elsewhere, another possibility that remains to be investigated is that the products of the reactions catalyzed by these enzymes are required for processes other than scent emission in the flowers. Indeed, it has been found that the flowers of many species accumulate glycosides of scent compounds as they age (Oka et al., 1999). Such non-volatile glycosides are also sometimes found in buds, and were therefore originally hypothesized to be obligatory "scent precursors." However, closer examination has shown that, in most cases, an increase in emission of a particular volatile is not accompanied by a corresponding decrease in levels of the glycoside of this volatile, as would be predicted by this hypothesis (Oka et al., 1999). The increased synthesis of such glycosides as the flowers age may account for the cessation of scent emission, although the specific roles of such glycosides in the flower remain to be determined.

Expression of genes encoding scent-biosynthetic enzymes in the C. breweri flower is temporally and spatially regulated during flower development. The mRNAs encoding LIS, IEMT, and BEAT are first detected in petal cells just before the flower opens, and their levels increase until they peak at or around anthesis and then begin to decline (Dudareva et al., 1996, 1998a; Wang et al., 1997). For all of these three genes, peak levels of the mRNAs occur 1 to 2 d ahead of the peaks of enzyme activity and emission of the corresponding compound. Similar results were found for these mRNAs in other parts of the flower.

Overall, our data show that a good positive correlation exists between the amount of mRNA, the amount of protein and enzymatic activity for each of these enzymes, and emission of the corresponding component up to the second or third d post anthesis. But beyond that point, the levels of scent enzymes remain relatively high despite declining levels of the corresponding mRNA and also without the concomitant emission of volatiles (Dudareva et al., 1996, 1999). These results also indicate that in C. breweri flowers, scent compounds are synthesized de novo in the epidermal cells of organs from which they are emitted (primarily the petals). Thus, the levels of activity of enzymes involved in scent production and, indirectly, scent emission are regulated mainly at the mRNA levels at the site of emission.

EVOLUTION OF FLORAL SCENT

An intriguing observation concerning floral scent is how variable this trait is. In many taxa, there are scented species that are closely related to non-scented ones, leading to the inescapable conclusion that the ability to produce and emit floral scent is an easily acquired, and easily lost, trait (Dudareva et al., 1996). Moreover, considering that floral scent is a complex mixture of chemicals, and practically no two closely related species emit identical mixtures of volatiles, it is clear that the ability to produce a specific floral scent volatile is an easily evolved trait. What is the basis for these evolutionary changes?

True to the complex nature of scent itself, the answer to the question posed above turns out to be complex as well. We have found that the genome of a close relative of C. breweri and its likely progenitor, the scentless Clarkia concinna, also contains the gene for linalool synthase (Cseke et al., 1998). Moreover, C. concinna flowers express LIS, but only in the stigma, and at such low levels that little linalool is produced and practically none is emitted (Raguso and Pichersky, 1995; Dudareva et al., 1996). Comparison of the promoter sequences of LIS from C. breweri and C. concinna revealed that although the promoter region in this gene from both species is almost identical, several insertions occur within the C. concinna promoter region relative to the C. breweri LIS promoter, two of them immediately upstream of the putative TATA box and a third inside the CAATT box (Cseke et al., 1998). It is possible that these differences in the LIS promoters are responsible for the vastly different expression characteristics of LIS in these two species, but this remains to be verified experimentally.

It appears that the difference between a linalool-emitting plant and a linalool non-emitter is not in the possession of a LIS gene per se, but in the mode of its regulation. In fact, the ability to synthesize linalool either in vegetative or floral parts is widespread in the plant kingdom, and has often been shown to be part of a defense response (as in maize, soybean, and possibly in the stigma of C. concinna), as well as being present in the scent of numerous flowers. It is therefore likely that virtually all plant genomes contain a LIS gene, and by a few changes in the regulation of expression of this gene, a plant may be able to easily recruit it for scent production.

A somewhat different scenario explains the ability of C. breweri flowers to produce methyleugenol and isomethyleugenol. We were able to show (Wang and Pichersky, 1999) that the IEMT gene, encoding the enzyme that methylates eugenol and isoeugenol, evolved recently from a gene encoding caffeic acid methyltransferase (COMT), which catalyzes the formation of ferulic acid and 5-hydroxyferulic acid, both non-volatile intermediates in the biosynthesis of lignin. The two enzymes are very similar, and changing only five to seven residues in one of them can convert the enzyme to preferentially act on the preferred substrates of the other (Wang and Pichersky, 1999). COMT is a ubiquitous gene throughout the plant kingdom, whereas IEMT most likely arose within Onagraceae, the family in which C. breweri is placed. Thus, unlike LIS, the ability of C. breweri to make the methyleugenol and isomethyleugenol components of its floral scent is due to the recent evolution of a new gene. However, like LIS, some C. breweri plants that do not emit (iso)methyleugenol have the IEMT gene in their genome but do not express it, indicating that the regulation of gene expression is also a factor in (iso)methyleugenol emission (Wang et al., 1997).

Non-scented C. concinna also contains genes with high sequence identify to C. breweri BEAT, but its flowers have little BEAT enzymatic activity. This is due first to the fact that the levels of properly processed transcripts of these genes in flowers are very low. In addition, the enzymes encoded by such transcripts have higher affinity with substrates other than benzylalcohol (e.g. heptanol; Nam et al., 1999). (Enzymes that can react with more than one substrate are not uncommon in secondary metabolism; see Wang and Pichersky, 1999.) It has recently been shown that BEAT is a member of a large family of plant acyltransferases (Dudareva et al., 1998; St.-Pierre et al., 1998). A few enzymes belonging to this family that catalyze the formation of various non-volatile (with the exception of benzylacetate) secondary compounds (e.g. anthocyanins and phytoalexins) have been identified. However, based on the presence in the Arabidopsis genome of perhaps 70 such related genes, it is clear that the functions of the majority of them remain to be determined. Such a large family represents a rich source of enzymes that may be involved in scent biosynthesis or serve as a pool for the evolution of new scent enzymes.

The methyltransferases SAMT and BAMT also belong to a newly identified family of enzymes. Although Arabidopsis has perhaps as many as 30 related genes of this family, and although several ESTs that are derived from genes of this family have been reported from Arabidopsis and other species, the reactions catalyzed by none of these proteins (with the exception of SAMT and BAMT) have been determined. This family is also likely to include scent enzymes such as the ones catalyzing the formation of methyljasmonate and methylcinnamate.

FUTURE PROSPECTS

It is clear from the above discussion that a major priority of scent research should be the continuation of the elucidation of biochemical pathways leading to scent biosynthesis and the identification and characterization of enzymes and genes controlling these pathways. In addition, the subcellular location of the synthesis of most of the scent compounds still needs to be determined, as well as the mechanisms controlling developmental changes and circadian rhythms of the pathways. On the evolutionary scale, it would be instructive to examine the molecular processes that bring about the variability in floral scent characteristics among different species, whether they are on the level of gene regulation, post-transcriptional regulation, or protein evolution. Finally, the contribution of specific scent compounds in attracting specific pollinators needs to be rigorously examined. The availability of scent genes should allow us to create transgenic lines whose floral bouquets differ by a single component, thus simplifying this task.

MOLECULAR PHYSIOLOGY OF PLANT NITROGEN USE EFFICIENCY AND BIOTECHNOLOGICAL OPTIONS FOR ITS ENHANCEMENT

Nitrogen use efficiency (NUE) in plants is a complex phenomenon that depends on a number of internal and external factors, which include soil nitrogen availability, its uptake and assimilation, photosynthetic carbon and reductant supply, carbon-nitrogen flux, nitrate signalling and regulation by light and hormones, to name a few.

The molecular basis for organism-wide regulation of nitrate assimilation is not yet fully understood, and biotechnological interventions to improve crop NUE have met with limited success so far. This article summarizes the physiological, biochemical and molecular aspects of NUE, QTL mapping studies as well as transgenic efforts to improve it in crop plants and model plants.

It encompasses primary and secondary N-assimilatory pathways and their interplay with carbon metabolism, as well as signalling and regulatory components outside the metabolic cascade. The article highlights the need for an integrated approach combining fertilizer management techniques with biotechnological interventions to improve N flux and NUE for Indian crop cultivars.

NITROGEN (N) is one of the most critical inputs that define crop productivity and yield under field conditions, and must be supplemented to meet the food production demands of an ever-increasing population. Efficient utilization of fertilizer N is essential to ensure better value for investment as well as to minimize the adverse impacts of the accumulation of reactive N species in the environment. The current average nitrogen use efficiency (NUE) in the field is approximately 33% and a substantial proportion of the remaining 67% is lost into the environment, especially in the intensively cropped areas. This concern was reflected in the recent Nanjing Declaration of the International Nitrogen Initiative, which called for immediate development of a comprehensive approach to optimize N management in every sphere of life.

Though the form and amount of N available to the plant can be improved by managing fertilizer-soil-water- air interactions, the innate efficiency of the plant to utilize this available N has to be tackled biologically. The biological processes involved include nitrogen uptake, translocation and assimilation, and their optimal contribution towards a desirable agricultural outcome, such as biomass growth and/or increased grain/leaf/ flower/fruit/ seed output, depending on the plant/crop involved. Identification of appropriate phenotypes, genotypes, molecular markers and target candidates for improvement of NUE poses a formidable challenge.

The purpose of this article is to summarize the current state of our understanding of the physiological and molecular aspects of plant N response and NUE, with a brief overview of the attempts made so far towards manipulating it and the possible options and strategies for future interventions.

Concept and Definition of NUE

As a concept, NUE includes N uptake, utilization or acquisition efficiency, expressed as a ratio of output (total plant N, grain N, biomass yield, grain yield) and input (total N, soil N or N-fertilizer applied). NUE is quantified based on apparent nitrogen recovery using physiological and agronomic parameters. Agronomic efficiency is an integrative index of total economic outputs relative to the available soil N (native and applied). Apparent nitrogen recovery is related to the efficiency of N uptake; physiological NUE deals with N utilization to produce grain or total plant dry matter. The most suitable way to estimate NUE depends on the crop, its harvest product and the processes involved in it.

Molecular Physiology of Nitrogen Uptake and Assimilation

Among the various forms of N available to the plant, nitrate (NO-3) is the most preferred source for most plants. It is taken up by active transport through the roots, distributed through the xylem and assimilated by the sequential action of the enzymes nitrate reductase (NR) and nitrite reductase (NiR). The end-product, ammonium (NH+4), is incorporated into amino acids via the glutamine synthetase (GS) and glutamate synthase (GOGAT) cycle.

The availability of organic acids is critical for the supply of carbon skeletons needed for amino acid synthesis, which in turn demands optimum partitioning of photosynthetic sugars and the coordinated operation of multiple metabolic and regulatory pathways.

Nitrate Uptake

Optimum uptake of nitrate is the first step to enhance N use in any plant. It has been established from a number of physiological studies that plants acquire their nitrate from the soil through the combined activities of a set of highand low-affinity transporter systems, with the influx of NO-3 being driven by the H+ gradient across the plasma membrane. Some of these transporters are constitutively expressed, while others are nitrate-inducible and subject to negative feedback regulation by the products of nitrate assimilation.

The low affinity transport system (LATS) is used preferentially at high external nitrate concentrations above 1 mM, while the high affinity transport system (HATS) works at low concentrations (1 ?M-1 mM).

LATS is constitutive in nature and possibly has a signalling role to induce the expression of HATS and nitrate assimilatory genes, presumably playing a nutritional role only above a certain threshold. The identification of two gene families, NRT1 and NRT2, on the basis of their deduced amino acid sequences has contributed towards unravelling the mechanisms of nitrate uptake in higher plants.

Physiology of Nitrate Reduction in Crops

A portion of the nitrate taken up is utilized/stored in the root cells, while the rest is transported to other parts of the plant. Due to the abundant availability of photosynthetic reductants, leaf mesophyll cells are the main sites of nitrate reduction. This is initiated by the NAD/NADP-dependent NR enzyme, which catalyses the two-electron reduction of nitrate to nitrite in the cytosol. Nitrite is transported into the chloroplast, where it is further reduced into ammonium ion by a ferredoxin-dependent NiR. Being the first, irreversible and often rate-determining step of the N-assimilatory pathway, nitrate reduction has been a favourite step for physiological and biochemical approaches to optimize fertilizer N use. The transgenic approaches have been dealt with separately later in the article.

Pattern of NR Activity

NR activity in leaf blades, expressed either as seasonal average or converted into seasonal input of reduced N, has been related to total reduced N, grain N and grain yield of cereals. The pattern of nitrate assimilation from different plant parts, viz. the main shoot of wheat, developing ear of wheat plants grown at different soil N levels and in the leaf blades at different stages of growth has revealed a direct positive correlation between increasing NR activity and increasing rates of nitrogenous fertilization. Most plant tissues have the capacity to assimilate nitrate, though their NR activity varies widely. Analysis of the shoot components revealed that leaf blades are the main sites in cereals like wheat, corn and barley. Detectable level of NR activity has also been observed in the developing ear components of wheat, barley and pod covers of chickpea. Among the ear components, the in vivo activity was highest in awns. It was also observed that the ontogenetical pattern of NR activity corresponds to the nitrate content, with which it is significantly correlated. The light/dark conditions also affect NR activity; heterotrophic nitrate assimilation in darkness is closely linked to the oxidative pentose phosphate pathway and the supply of glucose-6-phosphate. Under photoautotrophic conditions, glucose-6-phosphate dehydrogenase is inhibited by reduction with thioredoxin in light, thus replacing the heterotrophic dark nitrate assimilatory pathway with regulatory reactions functioning in light. These studies as well as bioenergetic calculations have indicated that both yield and N harvest or protein can be increased to some extent with adequate nitrogen supply by altered management practices, thus improving the fertilizer NUE.

Genotypic Differences in NR Activity

Genotypic differences in the NR levels have been reported in corn, wheat, sorghum and barley. In sorghum, a positive relationship between decline in the height of the plant and enhancement of NR activity was observed, though no such relationship was evident in tall and dwarf cultivars of wheat, T. aestivum. Wheat genotypes revealed over twofold variability in NR activity, which supports genetic findings that the enzyme level is highly heritable, its differences are reflected in N harvest and that hybrids could be bred with predictable NR levels by selecting parents appropriately. In the high NR genotypes, higher levels of NR activity were found under low N levels, often with significantly higher N concentration in the grains. They also have sustained activity at later stages of growth, such as flag leaf emergence and anthesis. The reasons for these genetic differences are not fully understood, except that the regulation operated at the level of gene expression and that low levels of NADH might limit NR activity in low NR genotypes.

Ammonium Assimilation

Ammonium is taken up directly through the roots, though uptake can also occur in a biphasic manner, involving LATS and HATS. Ammonium generated from primary nitrate assimilation, re-assimilation of internal metabolites or other secondary sources, is incorporated into amino acids in a reaction catalysed by GS and then by GOGAT. GS is the central enzyme in ammonium assimilation in plants with a cytosolic (GS1) and a plastidic (GS2) isoform. Similarly, GOGAT has two isoforms, a ferredoxindependent plastidic isoform (Fd-GOGAT) and a NADHdependent cytosolic isoform (NADH-GOGAT). The plastidic isoforms of both the enzymes (GS2 and Fd-GOGAT) are involved in primary ammonia assimilation, while their cytosolic isoforms are involved in secondary assimilation.

Secondary Ammonium Assimilation/remobilization

As nitrogen is a major limiting factor for plant growth, the efficient re-assimilation of metabolically generated ammonium is highly important for NUE, plant performance and to prevent loss of ammonia to the atmosphere. Cytosolic GS1 and NADH-GOGAT are of critical importance in this regard. GS1 has been proposed as a key component of NUE in plants and its metabolic role is particularly important for nitrogen remobilization and recycling in woody plants. Ammonium ions are also derived from the mitochondrial glycine decarboxylase reaction, which is an integral part of the photorespiratory carbon and nitrogen pathways. The importance of glutamate dehydrogenase (GDH) in higher plant N metabolism is still controversial, as it has never been clearly demonstrated that it plays a significant role either in ammonium assimilation or carbon recycling in plants. The role of GDH in N management and recycling has been recently reviewed in a number of whole-plant physiological studies performed on tobacco and maize.

Signalling and Regulation of Nitrogen Assimilation

Since N demand and its actual availability tend to vary in time, space and environmental conditions, the regulation of plant-N metabolism must be responsive to nutritional, metabolic and environmental cues. The following sections deal with the recent advances in our knowledge of the complex web of interactions in the regulation of nitrate assimilation by internal and external signals and its coordination with the overall metabolism of the plant.

Role of Nitrate

Nitrate is not only a nutrient but also a signal for the regulation of hundreds of nitrate-responsive genes, which include N and C metabolizing enzymes, redox enzymes and a whole range of signalling proteins and transcription factors. However, the mechanism of nitrate signalling is not well understood, though calcium and protein kinases have been implicated, as reviewed reccently. The transcriptional regulation of nitrate responsive genes could involve cis-acting regulatory sequences or nitrate response elements (NRE). Our recent genome-wide computational analysis of a previously reported NRE comprising the consensus sequence [(a/t)Ac/gTCA] based on NR and NiR turned out to be neither unique nor common to all nitrate responsive genes in Arabidopsis and rice, necessitating a fresh search for newer candidate NRE sequences. However, the identification of putative cis elements that are responsive to C/N signalling interactions indicates the possible combinatorial role of different cis-regulatory elements. Identification of such regulatory elements might provide an end-point for nitrate signalling and open up avenues for characterizing/manipulating the rest of the signalling pathway to enhance NUE.

Role of Light

Light is an additional signal that regulates the expression of many nitrate responsive genes. Light regulation of NR expression and activity operates differently in green plants and etiolated seedlings, and is mediated by different photoreceptors. The effects of light in green plants are probably mediated more indirectly through photosynthesis and sugars. Using pharmacological approaches, the phytochrome-mediated regulation of NR gene expression in etiolated maize seedlings was suggested to be mediated through G-protein, PI cycle and protein kinase C. Similar effects of cholera toxin and lithium ions were also found recently on NR-mRNA and activity in light-grown-dark-adapted rice seedlings, but both PMA and okadaic acid had inhibitory effects on NR-mRNA and activity, indicating different responses in maize and rice, either due to etiolated/green plant or C3/C4 differences.

Role of 14-3-3 Proteins

It is becoming increasingly evident that 14-3-3 proteins play a crucial role in bringing about metabolic coordination between enzymes of C and N metabolism, modulating their activity by binding them in a phosphorylationdependent manner.

NR, GS, sucrose-phosphate synthase (SPS), trehalose-phosphate synthase, glutamyl-tRNA synthetase, and an enzyme of folate metabolism have been found to bind to 14-3-3s in a phosphorylation-dependent manner. Experiments in transgenic potato plants indicate that repression of 14-3-3 proteins led to significant increases in NR and SPS activities.

More recently, the effect of repression of 14-3-3 genes on actual activity of NR in Nicotiana benthamiana leaves, was studied by silencing the Nb14-3-3a and Nb14-3-3b genes using virusinduced gene silencing method, which implicated Nb14-3-3a and/or Nb14-3-3b proteins in the inactivation of NR activity under darkness in N. benthamiana. The 14-3-3 proteins also interact with components of plant signalling pathways as observed in their interaction with RGS3, a negative regulator of the G-alpha subunits of heterotrimeric G proteins, suggesting a possible role in the regulation of G-protein signalling pathways, which in turn have been implicated in mediating light regulation of NR.

Role of Downstream N Metabolites

Nitrite accumulation is toxic to the plant and is also inhibitory to nitrate induction, whereas the effects of ammonium and glutamine vary depending on the tissue and plant type as well as conditions of the study.

The addition of ammonium or nitrate to N-limited whole plants or plant cells induces (at the transcript and/or activity level) enzymes of glycolysis and the Krebs cycle, which are required for the synthesis of 2-OG. Glutamate and 2-oxoglutarate have recently been shown to stimulate nitrate induction of NR and NiR in rice seedlings.

The role of glutamate as a signalling molecule in plant nitrogen metabolism has been reviewed recently, indicating that the manipulation of N nutrition leads to dynamic alterations in plant respiratory metabolism in response to changes in cellular energetic demands. Therefore, the roles and interactions of downstream N metabolites may have to be factored into strategies for optimizing N response and NUE.

Role of Hormones

Plant hormones like cytokinin have been shown to mimic the N-dependent regulation of gene expression in photosynthesis, cell cycling and translational machinery; hinting at a possible role in communicating the availability of nitrogen from roots to leaves. Additionally, N sensing and response also seem to be affected by the crosstalk between various plant hormones. Auxin synergistically affects cytokinin activity on cell division and organogenesis, while ABA antagonizes the cytokinin-mediated nitrogen signalling by negatively regulating cytokinininducible response regulator genes. Unlike cytokinins, which are positively regulated by nitrate, ABA biosynthesis is down regulated by nitrogen sufficiency. Although gibberellins do not seem to play any role in the control of nitrate assimilation, at least in the vegetative stages of Arabidopsis, benzyladenine in combination with nitrate was shown to enhance NR-specific mRNA. Despite these findings, establishing the role of hormones in nitrogen signalling needs further characterization of the complete signalling pathway.

Interaction of Nitrogen and Carbon Metabolism

The tight regulation of C/N metabolism has been revealed through numerous studies, which have indicated that the net photosynthesis rate and amount of photosynthetic components are correlated with the leaf-N content. The relative abundance of N pools in the plant plays a significant role in regulating the C/N metabolism. Nitrate supply has been shown to result in the decrease of starch synthesis and diversion of carbon towards the conversion of organic acids into amino acids. On the other hand, nitrate deficiency results in the decrease of many amino and organic acids, along with an increase in the level of several carbohydrates, phosphoesters and a handful of secondary metabolites. Recent studies on global gene expression have revealed that a significant number of the previously reported nitrate responsive genes actually required the presence of both nitrogen and sugar, suggesting significant interaction between C and N metabolites in regulating gene expression, with carbon modulating the effects of nitrogen and vice versa. Recent evidences of post-transcriptional control of C/N regulation by microRNAs have revealed globally coordinated regulation of specific sets of molecular machines in the plant cell.

Given the strong relationship between N and photosynthetic rates, plants maximize photosynthesis by optimizing partioning of N, which further depends on other environmental factors such as irradiance, nutrients, CO_2 concentration, etc. Consequently, the photosynthetic nitrogen use efficiency (PNUE) is determined by the rate of carbon assimilation per unit leaf nitrogen. Plants possessing C_4 photosynthesis have a greater PNUE than C_3 plants, owing to the C_4 concentrating mechanism that leads to CO_2 saturation of Rubisco. Higher CO_2 concentrations both compensate for the poor affinity of Rubisco for CO_2 and suppress oxygenase activity, consequently increasing the PNUE at elevated concentrations. Further evaluation of the key components of photosynthesis and interactions of C/N metabolites might offer avenues for improving N uilization by optmizing N content in accordance with photosynthetic demand.

Interaction of Nitrogen and Sulphur Metabolism

The importance of sulphur as a nutrient and its management vis-à-vis other nutrients like nitrate have been reviewed recently. Under sulphur-deficit conditions, reduced protein synthesis is accompanied by accumulation of organic and inorganic nitrogenous compounds, leading to reduced NUE, indicating the need to achieve optimum N/S balance for improved NUE.

Options for Improvement in NUE

NUE can be improved to some extent by optimizing fertilizer- soil-water interactions, though the biological aspects of crop improvement form the main purpose of this article. Regardless of the approach adopted, the challenges in improving NUE include optimization of N supply and demand, maximization of crop N uptake and assimilation, minimization of N losses and ultimately, specific improvements in the yields of biomass, leaves, fruits or grains, as the case may be. The current section deals with some of these approaches and their impact on crop NUE.

Fertilizer-N Application Management (repeated N fertilization in split doses)

Prevalent fertilizer management practices result in high nitrate content and NR activity in the first-formed leaf blades, which decline in the subsequently formed ones. The pattern was paralleled by soil nitrate concentration and its total content. Incubation of excised leaf blades in a nutrient solution containing 15 mM NO- 3 resulted in

slight increase in the NR activity of the lower leaf blade of wheat, while the activity of the upper ones was enhanced manifold; the level of enhancement being higher in 'high NR' cultivars than in 'low NR' cultivars. It has been demonstrated successfully that application of the same amount of nitrogen fertilizer in more than two splits under field conditions clearly increases the nitrogen availability at later stages of growth, exploiting the sub-optimal activity of the upper laminae in wheat.

Studies on Indian mustard genotypes with contrasting NUE showed that plants with high N uptake efficiency (UE) and high physiological N utilization efficiency (PUE) are able to not only take up N efficiently, but also utilize N efficiently. Such plants are highly desirable because they can be grown with limited N supply for environment-friendly farming systems. Genotypes with high UE accumulated higher N content than those with low UE under limited N conditions.

High PUE is essential for optimum seed yield, because these genotypes absorbed N efficiently. Although the genotype with high UE and low PUE takes up N efficiently from the soil, it remains unutilized in the form of non-protein-N, as UE showed significant positive correlation with the free amino acid pool.

Thus, development of such N-efficient genotypes, which can grow and yield well at low N levels further enhance options for better management of the applied N fertilizer.

Transgenic efforts to Manipulate NUE

The speed and precision of the transgenic approach enables one not only to test the candidate genes considered to be critical for NUE by overexpressing them, but also to identify such genes by knock-out mutations. The following sections describe various transgenic studies involving different categories of nitrate responsive genes.

Manipulation of Transporters

Studies in the last decade have shown that enhancing the uptake of N by overexpressing transporters may not necesarily improve NUE. For example, transgenic overexpression of a CHL1 cDNA (representing the constitutive HATS) driven by the cauliflower mosaic virus 35S promoter in a chl1 mutant, recovered the phenotype for the constitutive phase but not for the induced phase5. Similarly, the NO-3 contents in transgenic tobacco plants overexpressing the NpNRT2.1 gene (encoding HATS), were remarkably similar to their wild-type levels, despite an increase in the NO-3 influx. These findings indicate that genetic manipulation of nitrate uptake may not necessarily lead to concomitant improvement in nitrate retention, utilization or NUE, though it remains to be seen whether different plants respond differently to the overexpression of different transporters.

Manipulation NR and NiR Genes

NR has long been considered to be the rate-limiting step in nitrate assimilation. Efforts to improve NUE by manipulating NR and NiR genes have yielded mixed results, with transformed Nicotiana plumbaginifolia plants constitutively expressing NR, showing a temporarily delayed drought-induced loss in NR activity, thereby allowing more rapid recovery of N assimilation following shortterm water-deficit. At the transcriptional level, de-regulation of NR gene expression by constitutive expression in transgenic plants caused a reduction in nitrate levels in tissues of tobacco and potato. While factors such as NO-3 availability regulate flux through the pathway of N assimilation, the NR transformants were better equipped in terms of available

NR protein, which rapidly restored N assimilation. Though no tangible effects on biomass accumulation could be attributed in the short term, under field conditions of fluctuating water availability, constitutive NR expression was able to confer a physiological advantage by preventing slowly reversible losses in N-assimilation capacity. Similarly, overexpressing NiR genes in Arabidopsis and tobacco resulted in increased NiR transcript levels but decreased enzyme activity levels, which were attributed to post-translational modifications. Therefore, the utility of transgenic overexpression of NR/NiR for major improvements of NUE remains uncertain, though the possibility that different crops respond differently cannot be ruled out yet.

Manipulation of GS2 and Fd-GOGAT Genes

Improvement in NUE via manipulation of plastidic GS2 and Fd-GOGAT genes has met with limited success. Transgenic tobacco plants with twofold overexpression of GS2 were shown to have an improved capacity for photorespiration and an increased tolerance to high-intensity light. On the other hand, transgenics with reduced amount of GS2 had a diminished capacity for photorespiration and were photoinhibited more severely by high-intensity light compared to control plants. Overexpression of GS2 has also been reported in rice and tobacco, with improved reassimilation of ammonia in tobacco. Studies on barley mutants with reduced Fd-GOGAT revealed changes in various nitrogenous metabolites, decreased leaf protein, Rubisco activity and nitrate content. While these studies hint at the potential of such transgenic attempts, most of them have been inconclusive regarding NUE so far, due to lack of physiological and agronomic data.

Manipulation of GS1 and NADH-GOGAT Genes

Ectopic expression of pea GS1 in tobacco leaves was suggested to provide an additional or an alternative route for the reassimilation of photorespiratory ammonium, resulting in an increase in the efficiency of N assimilation and enhanced plant growth. Efforts to raise more efficient GS1 transgenic lines have met with varying degrees of success-, with Man et al. providing additional empirical evidence for enhanced nitrogen-assimilation efficiency in GS1 transgenic lines. Transgenic overexpression and underexpression studies to modulate the expression of NADH-GOGAT in alfalfa and rice plants, have implicated the involvement of GS1 in the export of N via phloem in senescing leaves. On the other hand, in case of developing leaf blades and spikelets, NADHGOGAT was implicated in the utilization of glutamine transported from senescing organs. Though these genes of secondary ammonia assimilation appear to be more viable candidates for improving NUE, the degree of success needs to be tested across crops and cropping conditions.

Manipulation of Signalling Targets

Yanagisawa et al. generated transgenic Arabidopsis lines overexpressing Dof1, a maize protein that belongs to the Dof family of plant-specific transcription factors known to activate the expression of several C-metabolizing genes associated with organic acid metabolism. The transformants showed up to 30% higher N content, higher levels of amino acids, better growth under low-nitrogen conditions and higher levels of mRNAs and enzyme activities for PEP carboxylase and pyruvate kinase, without any reduction of NR, GS and GOGAT RNAs. The genes upregulated by Dof1 overexpression clearly belong to the list of known nitrate responsive genes, opening up attractive possibilities of improving NUE through coordinated expression of N and C metabolizing genes. A few other attempts to manipulate signalling/regulatory proteins have been made, without significant advantage in terms of NUE. Other attempts, such as the one to manipulate a MADS box protein that controls nitrate-induced changes in root architecture, have not been assessed for their impact on NUE.

Manipulation of Source-sink Relationships and Nutritional Quality

Molecular manipulation of certain key enzymes of N metabolism provides an attractive means to enhance the nutritional value of plant products along with increasing the quality and quantity of seed proteins in crop plants. The enzyme asparagine synthetase (AS) catalyses asparagine, one major function of which is to transport and store nitrogen according to the plant's need. It can also reallocate nitrogen during specific developmental stages and environmental changes. The efficiency of protein synthesis has been shown to be dependent on the light/ dark regulation of AS activities, with elevation of leaf AS activities and Asn levels being used as parameters to screen for high-grain protein cultivars in maize and rye. On the other hand, regulating the expression of the ASN1 gene to manipulate the relationship between Asn and seed N status might enhance nutritional quality.

Studies have implicated AS as one of the major controlling forces for nitrogen flux when GS is limiting in plants, and several studies on transgenic overexpression of AS genes have revealed enhanced seed protein content and total protein content-. Recent genetic modification of rice and wheat using barley alanine amino transferase (AAT) gene have also yielded encouraging results, including increased biomass and seed yield compared to their wild-type counterparts. Recently, Arcadia Biosciences claimed to have improved NUE by transgenic overexpression of AAT in canola, Arabidopsis, tobacco and rice, though their actual field performance is yet to be ascertained. In other studies, two potentially important N regulation systems of Arabidopsis, PII and GCN2 have been targetted; though detailed analyses of the effect of their transgenic overexpression on NUE are yet to be reported.

QTL Mapping to Find New Targets for Manipulation

The development of molecular markers has facilitated the evaluation of the inheritance of NUE using specific quantitative trait loci (QTLs) that could be identified. QTLs for NUE have been identified in mapping populations of maize, rice, barley and Arabidopsis, and their association with plant N status has been reviewed recently. In maize, studies on different genotypes or populations of recombinant inbred lines based on NUE components, chromosomal regions and putative candidate genes have hinted at some factors that might control yield and its components directly or indirectly, when the amount of N fertilizers provided to the plant is varied.

Future Perspectives

It is clearly evident that optimizing the plants NUE goes beyond the primary process of uptake and reduction of nitrate, involving a paraphernalia of events, including metabolite partitioning, secondary remobilization, C-N interactions, as well as signalling pathways and regulatory controls outside the metabolic cascades. Despite the various attempts to manipulate each of the above steps in some plant or the other, we are far from finding a universal switch that controls NUE in all plants. However, transgenic studies and QTL approaches seem to increasingly suggest that the enzymes of secondary ammonia remobilization are better targets for manipulation, followed by regulatory processes that control N-C flux, rather than the individual genes/enzymes of primary nitrate assimilation. However, it is possible that different plants respond differently to various targets of manipulating NUE, especially since field-level improvements in NUE result from many more complex interactions. The need of the hour is integration of physiology and molecular genetics involving Indian crop cultivars to optimize yields in different genotyopes and environmental conditions.

12

Evolvability of Physiological and Biochemical Traits

A longstanding challenge for biologists has been to explain not just how organisms are adapted to diverse environments, but how these adaptations arise. Although natural selection is clearly sufficient to act on heritable variation, is this heritable variation sufficient to yield complex adaptations and how does this variation itself arise? Much prior focus has been on mutation of single nucleotides in genes. This process is common and can have dramatic phenotypes, but could be limited in its ability to culminate in complex adaptations for two kinds of reasons: (i) because natural selection is powerful, it can purge genetic variation, and (ii) evolutionary transition from the absence to the presence of a complex adaptation seemingly requires multiple mutations at the right place and time and in the right sequence, with each intermediate stage having increased overall fitness; this seems highly improbable.

Because the networks that organisms comprise are hierarchical and redundant and have modular structure, however, single-nucleotide mutations can have large and tolerable impacts. Diverse mechanisms, collectively evolutionary capacitors, can shield genetic variation from the purgative of selection. These features can enable evolution to proceed via single-nucleotide mutation. Importantly, single-nucleotide mutation usually only modifies existing genes rather than creating new ones, and numerous other mechanisms eclipse single-nucleotide mutation in creating genetic variation. These include gene duplication (both segmental and whole-genome), lateral gene transfer, hybridization, mobile genetic elements and symbiosis. Other processes can scramble and reassemble nucleotide sequence. The mechanisms beyond single-gene mutation offer considerable promise in detailing the evolution of complex physiological and biochemical traits, and have already done so for several morphological traits.

INTRODUCTION

Despite the vulnerability and limitations of life processes, organisms exploit seemingly inhospitable environments and perform remarkable feats. The biochemical and physiological mechanisms that underlie these extraordinary capacities have been a major focus of experimental biologists, increasingly using the toolkits of molecular biology and genome-enabled research. Indeed, such research is increasingly revealing the inner workings of these mechanisms, as many of the other contributions to this review volume demonstrate. But how have these mechanisms themselves come into being?

Many biologists, if pressed, invoke 'adaptation by natural selection' as an answer to this question. For some biologists, the answer stops there. How else could the extraordinary biochemical-physiological traits that so obviously enable organisms to function in diverse environments and are so clearly 'tuned' to each organism's environment have arisen except by natural selection? Consolidating this assessment are numerous examples of evolutionary convergence and parallelism, wherein functionally equivalent or identical mechanisms have arisen multiple times in response to similar selective pressures.

Other biologists will go further, and emphasize consistency with the fundamental principles of natural selection: (a) individuals in a population vary in genotype; (b) individuals in a population consequently vary in phenotype (including biochemical and physiological traits); (c) individuals in a population consequently vary in fitness; i.e. reproductive output. Hence genotypes that encode higher fitness will come to predominate in a population. Documentation of points (a) and (b) is now abundant both within and among populations, species and higher taxa, and documentation of point (c) in natural populations is growing (Endler, 1986; Feder et al., 2000; Feder and Mitchell-Olds, 2003; Van Straalen and Roelofs, 2006). Four examples of diverse vintage are considered below.

(1) In human populations (Ingram, 1963; Williams et al., 2005), mutation of nucleotide 17 from A to G in the sequence encoding the beta-subunits of the Hemoglobin A protein converts a glutamine to a valine. As a result, hemoglobins form fibrils via interactions due to the valine. Consequently, the shape of the hemoglobin-containing erythrocyte changes. Individuals homozygous for the mutation develop sickle-cell disease, and have lower fitness. Individuals heterozygous for the mutation, by contrast, have enhanced resistance to malaria (Plasmodium infection) by some as-yet-unknown mechanism, which might include an increased immune response, filtering of infected erythrocytes by the spleen, and/or outright inhospitability to the Plasmodium. This is a classic example of a balanced polymorphism, wherein the advantages to the heterozygote offset the disadvantages to the homozygote mutant.

(2) The fish Fundulus heteroclitus inhabits brackish water habitats along the eastern coast of North America, and so experiences enormous variation in temperature. A key component of the adaptation to this thermal gradient has been evolution of the ldh-b gene, encoding one component of the enzyme lactate dehydrogenase (LDH). Specifically, northern populations express more of this gene's product than southern populations, which may compensate for the cooler average temperatures of the northern populations. Evolved differences between alleles are both in protein function (i.e. in coding sequence) and in gene expression (i.e. in the proximal promoter region). Indeed, as initially reported (Segal et al., 1996), the north-south variation in proximal promoter sequence is sufficient for corresponding differences in gene expression.

(3) Beach mice (Peromyscus) inhabit beaches with diverse sand coloration, and mismatches of fur and sand color

enhance predation on mice (Hoekstra et al., 2006). Thus mouse populations on dark substrates have repeatedly evolved melanic coloration. In one case, a single-nucleotide variant (changing arginine 65 to cysteine) of the melanocortin-1-receptor gene, Mc1r, is strongly associated with melanism; this gene's product is key to the production of dark-pigmented fur. In other cases, no such genetic variation is evident, suggesting that melanism has arisen through at least two different genetic routes.

(4) Brewer's yeast (Saccharomyces cerevisiae) makes ethanol via the alcohol dehydrogenase enzyme ADH1, which it then feeds into the tricarboxylic acid (TCA) cycle via the enzyme ADH2. Thomson et al. (Thomson et al., 2005) have theorized that ancestral yeast had a single ADH that, much like vertebrate LDH and lactate, recycled NADH for use in glycolysis by producing, in this case, ethanol; ethanol could readily be lost to the environment before it poisoned the yeast. When fleshy fruits arose and Saccharomyces colonized them, yeast then evolved enhanced biosynthetic capacity for ethanol (and tolerance for it), which enabled yeast to triumph over microbial competitors in the fruit. If so ADH2, which is specialized for ethanol catabolism, would only have been needed post hoc. Using maximium likelihood techniques (Thornton, 2004), the authors inferred the nucleotide substitutions that must have transformed the common ancestor of the ADH1- and ADH2-encoding genes into their present state, synthesized the corresponding enzyme, and confirmed that it was catalytically specialized for ethanol synthesis. Importantly, this work signifies the possibility of reconstructing the sequence of single-nucleotide mutations that might transform one protein to another.

These are only four of many examples of such work (Hochachka and Somero, 2002; Watt and Dean, 2000), but are noteworthy for several reasons. First, each exemplifies an unambiguous linkage among gene, trait, whole-organism function or performance, and fitness. Second, each involves alternative single nucleotide polymorphisms (SNPs), alleles or haplotypes segregating in natural populations. This variation could plausibly have arisen through the processes of single-nucleotide mutation: replication error, or damage to a base due to chemical agents or ionizing radiation followed by imperfect repair. Plausibly, multiple rounds of such random mutagenesis could have produced even the more complex haplotypes that are segregating in the latter three examples. Indeed, SNPs are common, account for a substantial fraction of phenotypic variation within populations, and are the basis for the coming era of pharmacogenomics and personalized medicine. No less a personage than Ernst Mayr (Mayr and Provine, 1998), p. 20 stated that `...gene mutations are the only raw material of evolution' [as cited by Ryan (Ryan, 2006)], with which many biologists concur. But is single-nucleotide mutagenesis the only process generating genetic variation that is consequential for fitness in natural populations? And is this process sufficient to account for the remarkably complex adaptations of living things?

Is Simple Single-nucleotide Mutation Sufficient to Generate Complex Adaptations?

Although Charles Darwin understood neither heredity nor mutation in detail, he clearly anticipated that the origin of complex adaptations through the accumulation of small changes would challenge his theory (Darwin, 1859). Also well in advance of the genomic era, Richard Goldschmidt proposed that, because mutations of small effect would be insufficient to yield complex adaptations, rare mutations of large effect were needed (Goldschmidt, 1960). [As Ernst Mayr (Mayr, 1963) again pointed out, this proposition was itself implausible - and where were these mutations of large effect and how did they arise?] Even today, the `irreducibility of complexity' (Behe, 1996) is a major talking point in the creationism and `Intelligent Design' political campaigns against the science of evolution - alongside the absence of fossils documenting `missing links'. But if `genes are living fossils' (E. Gaucher, personal communication), the molecular paleontology of modern genomics and its successor sciences have now revealed numerous mechanisms that can create massive variation in genomes upon which evolution can act. First, however, what are the evolutionary issues that these new discoveries may resolve?

Evolvability

According to Falconer (Falconer, 1981), the change in the mean value of a trait from generation to generation is the product of two variables: the intensity of natural selection and the genetic variance of the trait. Intensity of selection is a variable readily comprehensible to comparative physiologists, and needing little substantiation. Nonetheless, Falconer's equation states that no matter how intense selection may be, evolutionary change cannot ensue unless adequate genetic variation is present; all else being equal, evolvability is proportional to genetic variability. Evolvability, however, is seemingly both an independent and a dependent variable. Ongoing selection can purge a population of genetic variability if the intensity of purifying selection exceeds the ability of mutation and other variational processes to generate it - unless, that is, something shields genetic variability from selection. How, then, can genetic variability either be generated de novo or shielded from the purgative of selection? And how can genetic variability be generated such that at least some of it is neither so minor as to impede evolution nor so extreme as to be deleterious?

The Upgrade Pathway

A mechanical engineer could, in principle, transform an automobile into a powered boat by adding a hull, rudder and propeller, and by making appropriate adjustments in the transmission and steering mechanisms. Transformation of the same engine into a machine for powered flight is likewise feasible (cf. the Wright brothers, 1902), or of a propeller-driven aircraft into a jet aircraft or rocket. A computer engineer could upgrade the performance of a computer by installing a new operating system, input/output devices, and firmware. In such cases, human engineers have luxuries unavailable to evolving organisms. First, the engineers can readily power down the machine they are transforming, make changes and restart it;

living things can perform the counterpart operation only with great difficulty (but see embryogenesis and metamorphosis). Second, engineers can obtain an extraordinary variety of alternative parts, or invent novel parts if necessary; if relying on simple mutation, evolving organisms can proceed only by modifying the genes already at hand (Long et al., 2003).

Third, an engineer can conceive that, for example, to transform an automobile into a powered boat, requires simultaneous changes in the exoskeleton, propulsive system, steering system, etc., and implement these changes simultaneously. If relying on simple mutation, evolving organisms must either await numerous successive mutations, none of which enhances fitness without the others, or an incredibly improbable bout of massive simultaneous mutation. Weinreich et al. (Weinreich et al., 2006) addressed a relevant situation experimentally: bacteria in which five mutations confer maximal resistance against an antibiotic. These five mutations could occur in any of 5! (120) possible orders, of which only 18 successively increase antibiotic resistance.

The remainder are prohibited because at least one intermediate stage decreases antibiotic resistance. Finally, an engineer can discard failed attempts and begin again, whereas organisms in which mutation decreases fitness substantially are conceivably doomed evolutionarily. How, then, does a flightless organism evolve flight, a terrestrial organism an aquatic existence, or any less-adapted organism the extraordinary features that comparative physiologists have discovered?

Network Properties Enable Evolvability

The flowering of interactomics, network biology, and systems biology has revealed how single-nucleotide substitutions (and/or much larger mutations; see below) may in principle yield transformations of large effect, and thereby create adaptation. Although network biology is still in its infancy, several points are clear. Like any system, biological systems can be characterized as networks of interacting components, be they genes and transcription factors, messengers and signalling pathways, or species in a biological community (Barabâasi, 2003). Not all components in a biological network are equal; some are relatively highly connected (so-called `hubs') and some less so, with the number of connections obeying a power-law distribution. This `scale-free' pattern differs from random networks, in which the number of connections is a Poisson distribution. Not all links among components are of equal strength (Csermely, 2006). Finally, the structure of the network is consequential for its function and its resistance to perturbation (i.e. robustness). While a more extensive general discussion of networks is impossible here (but see, for example, Barabaasi and Csermely), several features are critical to evolvability.

Biological Networks are Hierarchical

Because all components in a network are connected, perturbing one component can perturb many - especially if that component is a hub. Examples are numerous: single-gene knock-outs can be lethal, as can pharmacological inhibition of a single enzyme or signaling molecule, and removal of a single `keystone species' can devastate a natural community. More benignly, a small number of transcription factor or signaling molecules can create a cascade of impacts downstream that effect growth, development, and regulation. Such organization may be inevitable for complex systems (Csete and Doyle, 2004): the necessity for integrating numerous complex systems requires `common currencies' such as ATP and the major second messengers.

Biological Networks are Modular

Networks are heterogeneous in the strength and number of connections. Some sub-organizations occur multiple times and are highly connected internally, but are somewhat interchangeable with their counterparts. These are modules, biological equivalents of the individual telephones in a telephone network, individual circuit boards in a computer or parts in a machine, or individual computers in a network. In both development and intracellular signaling, for example, specific modules repeatedly appear in diverse pathways (Carroll, 2005; Pereira-Leal et al., 2006 Pereira-Leal and Teichmann, 2005).

Biological Networks are Redundant and Robust

Networks are more like roller coasters (in which removal of an average single strut has little impact on function) than like houses of cards (in which removal of an average card causes the entire structure to collapse) (Feder, 1996). An exhaustive account of how, at every level of biological organization, change in or removal of components does not affect the performance of the whole has recently been published (Wagner, 2005). Alternative components whose substitution has no impact are said to occupy the same `functional space', and these functional spaces are remarkably extensive. For example, although some single-gene knockouts can be harmful or lethal, the vast majority of genes can be deleted with no apparent ill effect (Feder and Walser, 2005).

Combining these features yields a recipe for evolvability. Because of redundancy and robustness, mutation can tinker with numerous individual components without jeopardizing the whole. Because of modularity, a mutated module can be substituted for an original module (Carroll, 2005; Pereira-Leal et al., 2006; Pereira-Leal and Teichmann, 2005). Because of hierarchical organization, some small changes can trigger major changes downstream. Thus, while one or more single-nucleotide mutations in the average gene may seldom suffice to generate major adaptive novelty, such mutations in key genes in the right module at the right time may have enormous impacts. Recently there have been several reviews or similar theses advanced (Kirschner and Gerhart, 2005; Carroll, 2005; Carroll et al., 2005; Davidson and Erwin, 2006).

The most relevant formalization of this thinking is the duplication-degeneration/complementation model of Force, Lynch and colleagues, originally applied to duplicate genes (Force et al., 1999) but recently extended to networks in general (Force et al., 2005). It reasons that when a component is duplicated [and recall that many components and modules are already redundant], only one duplicate is necessary to maintain the pre-existing function. When the second component undergoes change, it may either degenerate (as selection need not maintain it in view of the duplicate) or take on a differing function (like the gene encoding ADH2, discussed above). The differing function can either be an entirely new one (neofunctionalization, as when change results in a gene's promoter acquiring a new response element), or the two duplicates can each take on part of the function of the original component (subfunctionalization, as exemplified by the genes encoding ADH1 and ADH2). Once refunctionalized, the former duplicate may then be preserved by selection.

Again because of hierarchical organization, refunctionalization of a single component can trigger major changes downstream. Perhaps the most notorious example of this final point concerns the homeobox or Hox genes, which have undergone duplication/refunctionalization in development and now specify whether a body segment grows a limb and/or wing and/or haltere and/or other structure, depending on which particular Hox paralog is active in that segment (Carroll et al., 2005).

The Hox paralogs have acquired the ability to respond appropriately for their position in the developing embryo via refunctionalization of their promoters to bind transcription factors that vary in concentration from anterior to posterior. Additional spectacular examples involve the origin of armor plating in stickleback fish (e.g. Colosimo et al., 2005) and coloration in insects (e.g. Gompel et al., 2005).

Interestingly, it has been suggested that the typical network structure of living things is itself a facilitator of evolvability, at least in theory (Oikonomou and Cluzel, 2006). These authors modeled the outputs of networks with equal numbers of components but with different topologies, random versus scale-free (i.e. resembling many biological networks), using Boolean threshold dynamics to compare the output with a target function. Emulating natural selection, they then randomly mutated the components of the two networks, and selected for or against mutants according to their resemblance to the target function. In this simulation, the scale-free networks evolved more rapidly and regularly toward the target function than did the random networks.

Evolutionary Capacitors Enable Evolvability

Duplication-degeneration/refunctionalization takes time and, as noted above, natural selection is fully able to purge temporarily unsuccessful intermediates from a population before conditions favor their preservation and/or subsequent evolution. Evolutionary capacitors are mechanisms that shield variation from the purgative of selection. The term was proposed by the editors of the journal Nature as the title of the first report that a molecular chaperone, Hsp90, might act in this way (Rutherford and Lindquist, 1998).

Molecular chaperones are proteins (or other compounds) that interact specifically with other `client' proteins, normally in the context of aggregation deterrence, protein folding and/or degradation; many chaperones are heat-shock or stress proteins. Hsp90 is a peculiar chaperone in that it is normally present and at relatively high concentrations in eukaryotic cells, where it interacts with numerous client proteins. It is thus not specific for a particular client. Hsp90 is also a typical chaperone in that it will bind proteins not in their native (i.e. fully and normally folded) state and assist in their proper folding. Rutherford and Lindquist hypothesized that, because of these properties, Hsp90 could bind mutant gene products that if unchaperoned would yield dramatic phenotypes - and the ensuing loss of their encoding genes due to selection.

Hsp90 would thus deter the elimination of these mutations. Upon environmental stress, however, the profusion of stress-damaged proteins in the cell would titrate Hsp90 away from these mutant gene products, allowing their phenotypes to emerge. Hypothetically, if these emergent novel phenotypes were beneficial, they could then be preserved by selection. Although no beneficial mutation has yet been discovered to have arisen by this mechanism in natural or experimental evolution, Lindquist and colleagues have now assembled numerous data consistent with most components of this hypothesis (Cowen and Lindquist, 2005; Queitsch et al., 2002; True et al., 2004), and their work has stimulated broad interest in evolutionary capacitors.

In hindsight, many features of living things are evolutionary capacitors or render genetic variation cryptic (Gibson and Dworkin, 2004; Masel, 2006). These include ploidy >1, recombination, segregation and sex. Like chaperones, these can protect gene products or combinations thereof from selection and suddenly reveal them. Likewise, chromatin configuration (Kornberg and Lorch, 1999; Segal et al., 2006), epigenetic modification (e.g. gene imprinting, methylation), alternative splicing and phenotypic plasticity in general, can all suppress/ reveal genetic variation. Alternative splicing is especially spectacular in this regard, as a single gene can encode >38 000 splice variants (Schmucker et al., 2000). Indeed, evidence is mounting that natural selection can manipulate splicing to express variants appropriate for a given environment (Marden, 2006). Finally, the intronic nature of eukaryotic genes provides an additional site for genetic variation. Mattick and colleagues have hypothesized that introns have enabled the evolution of encoding of RNAs that themselves do not encode proteins, and that these non-coding RNAs are in turn responsible for the complexity of eukaryotes (Mattick, 2003).

The Generation of Large-scale Mutation

Now that techniques are available that can assess and compare most genes in multiple genomes, we are recognizing that largescale (i.e. involving many nucleotides at once, if not entire genomes) mutations, in combination with network features and evolutionary capacitance, may play a correspondingly major role in the genesis of biological variation, especially the variations that comparative physiologists call `adaptations'. Unlike single-nucleotide mutation, which unarguably remains an important component of the evolutionary toolkit, these readily yield plausible scenarios of large-scale evolutionary change and origin of new features (Long et al., 2003); termed `genomic creativity' (Ryan, 2006).

Gene Multiplication

Duplication of Genes within the Genome

The standard genetic machinery of DNA replication, recombination and segregation is not perfect. Not only does this yield single-nucleotide mutations, but also duplications of larger fractions of the genome if not entire genomes.

There are several excellent introductions to the topic (Hurles, 2004; Zhang, 2003), which emphasize that gene duplications are about as frequent as single-nucleotide mutations. As a result 30-65% of the genes in eukaryotic genomes are duplicate [table 1 in Zhang (Zhang, 2003)]. Lynch and Conery (Lynch and Conery, 2000) estimated that one in a hundred genes is duplicated and fixed every million years. Thus, gene duplication is not a rare occurrence.

Duplications of nucleotide sequences arise primarily by two distinct mechanisms (Hurles, 2004; Zhang, 2003). In the first (tandem or segmental duplication), unequal recombination occurs between homologous sequences at two places in the genome (within the same chromatid, between sister chromatids, between different chromosomes).

In the second (retroposition), a transcribed mRNA sequence is reverse-transcribed, and the resultant DNA inserted into a chromosome.

Thus retroposition (a) can be into any accessible chromatin in the genome and is not limited to homologous sequence, (b) only duplicates a transcribed sequence, which contains no promoter sequence and need not contain introns, (c) duplicates genes or operons one at a time; in contrast, tandem duplications may be of sequences containing many genes and intergenic regions. Genome-wide studies reveal the extraordinary impact of these processes.

Our own species is remarkable for numerous large segmental duplications (Bailey and Eichler, 2006), as is evident in whole-genome displays. The Drosophila genome contains hundreds of genes lacking introns but otherwise close matches of intron-containing genes on other chromosomes (Betran et al., 2002; Emerson et al., 2004; Long et al., 2003), evidently the result of retroposition.

Errors in segregation during meiosis can also result in extra copies of entire chromosomes or, the utmost in gene duplication, of the entire genome. Indeed, whole genome duplication has clearly occurred in yeast and plants, and two rounds of duplication occurred in the chordate lineage (Dehal and Boore, 2005).

Introduction of Genes from Outside the Genome

Hybridization also has long been recognized as a potential source of genomic change. By definition, it involves the combination of two distinct genomes, typically due to the failure of reproductive isolative mechanisms. Although perhaps not an everyday occurrence, novel species have clearly arisen in nature as a result (reviewed by Ryan, 2006).

Much more frequently, species acquire new genes by lateral or horizontal gene transfer from another species, sometimes involving a biological vector and sometimes simply by uptake of DNA (Bushman, 2002). The likely magnitude of this transfer is such that it has been concluded that no real `tree of life' exists because all `branches' have interchanged genes and hence are genealogically indistinct (Doolittle, 1999). Again, the combination of genomic information and bioinformatic tools establishes how extensive this interchange is.

Finally, few species of eukaryotes are, strictly speaking, single species, but actually are multispecies consortia of symbiotes and their genomes (Margulis and Sagan, 2002. Whatever one's reaction to the more controversial aspects of Margulis's thesis, interactions of symbiont genomes and host genomes have clearly brought about the evolution of what physiologists classically regard as adaptations, such as the bacteriocyte of aphids (Moran and Degnan, 2006), the light organ of squids (Koropatnick et al., 2004), and the rhizosphere of plant roots (Marx, 2004).

Mobile Genetic Elements

Some DNA sequences encode the capacity to move within or among genomes. A first class of mobile element moves by encoding a mRNA that is reverse-transcribed and inserts elsewhere in the genome; these are aptly named retrotransposons. A second class either copies itself or excises from its original site and inserts elsewhere, but remains DNA throughout. Those unfamiliar with transposable elements tend vastly to underestimate their impact on evolutionary change (Kazazian, 2004); some (V. Walbot, personal communication; M. Evgen'ev, personal communication) attribute perhaps 40-90% of all evolutionary change to transposable elements.

The simplest impact of these elements is to interrupt the sequence of their current (and often former) site in the genome; i.e. insertional mutagenesis. Transposable elements sometimes capture portions of DNA sequence adjacent to their original insertion site and, when they excise, carry it with them. They may thus convey novel regulatory and/or coding sequence to a new insertion site, and thereby endow the new host gene with novel properties. Transposable elements now constitute large fractions of all eukaryotic genomes sequenced so far (e.g. approximately 50% in humans).

Scrambling Genes within Genomes

Once genomes have acquired new genes, whether by duplication or from another species, numerous processes in addition to single-nucleotide mutation may modify them, sometimes far more dramatically than can single-gene mutation. As noted, retroposition, segmental duplication and transposable elements can each transfer sequence within the genome, creating combinations of sequences that were previously non-existent.

Alternative splicing can likewise create novel combinations of gene products. Additionally, mechanisms such as exon shuffling, gene fission/fusion can create new genes (Long et al., 2003). In some rare cases, novel genes can even arise from nonsense or intergenic sequence (Long et al., 2003).

Interaction among these Features

The foregoing mechanisms, although often individually distinct, can interact to yield evolutionary novelty. For example (Zhang et al., 2004), the Drosophila gene jingwei arose when a first gene, yellow emperor, duplicated to yield a second gene, yande, and a third gene, alcohol dehydrogenase, retroposed into the third intron of yande. Thereafter, at least 30 non-silent single-nucleotide mutations yielded the present-day gene. Unlike its ADH ancestor, which is specialized to detoxify and assimilate ethanol, the substrate specificity of jingwei is for long-chain primary alcohols, which are important in hormone and pheromone metabolism.

In another example (Walser et al., 2006), the P element, a transposable element of Drosophila, needs physical access to decondensed DNA to insert itself. In many genes, the chromatin is normally condensed and in nucleosomes, except when the gene is actually being expressed (i.e. chromatin is often an evolutionary capacitor). Perhaps due to their role as an emergency response to stress, the proximal promoters of the genes encoding heat-shock proteins are constitutively decondensed and nucleosome-free. This has made the heat-shock promoters distinctively susceptible to the insertion of P elements, which selection may then maintain or eliminate to manipulate the expression of their host genes.

Finally, because the sequence of transposable elements is often highly conserved, their insertion throughout the genome creates opportunities for homologous recombination among distant sites, as noted above. Such recombination, in turn, may invert, duplicate or otherwise scramble genes, multi-genic nucleotide sequences, or segments of chromosomes.

Observing that one class of transposable element, a form of Alu element, is often present at the boundaries of segmental duplications in the human genome, Bailey and Eichler (Bailey et al., 2003; Liu et al., 2003) have hypothesized that the extensive segmental duplication in the human genome is due to the proliferation of one class of Alu elements in the human lineage (but not in close relatives), which then allowed ectopic recombination. Thus, to the extent that our species owes its capacities to gene duplication, these elements may have underlain the evolution of humankind.

CONCLUSION AND PROSPECTUS

The advent of molecular, genomic and bioinformatic techniques and their increasing applicability to diverse species has enormously enhanced experimental biologists' ability to understand `how animals work' (Schmidt-Nielsen, 1972). Adaptational biology will be incomplete, however, until the understanding of how adaptations came into being is equally advanced.

This understanding may well come about through sustained interaction with modern evolutionary biologists, evolutionary genomicists and evolutionary systems biologists. One clear outcome of such interaction is that single-nucleotide mutation, often the mainstay of adaptational biologists' evolutionary thinking, will become viewed as only one of several mechanisms in evolution's toolkit. The other mechanisms may be far more powerful than single-nucleotide mutation in facilitating evolvability and, although they have not done so yet, be able to explain in detail the origin of the complex traits that fascinate adaptational biologists.

13

Molecular Physiological Analysis

Arabidopsis (Arabidopsis thaliana) possesses two isoforms of plastidic ATP/ADP transporters (AtNTT1 and AtNTT2) exhibiting similar biochemical properties. To analyze the function of both isoforms on the molecular level, we examined the expression pattern of both genes by northern-blot analysis and promoter- -glucuronidase fusions. AtNTT1 represents a sugar-induced gene mainly expressed in stem and roots, whereas AtNTT2 is expressed in several Arabidopsis tissues with highest accumulation in developing roots and young cotyledons. Developing lipid-storing seeds hardly contained AtNTT1 or -2 transcripts. The absence of a functional AtNTT1 gene affected plant development only slightly, whereas AtNTT2::T-DNA, AtNTT1-2::T-DNA, and RNA interference (RNAi) plants showed retarded plant development, mainly characterized by a reduced ability to generate primary roots and a delayed chlorophyll accumulation in seedlings. Electron microscopic examination of chloroplast substructure also revealed an impaired formation of thylakoids in RNAi seedlings.

Moreover, RNAi- and AtNTT1-2::T-DNA plants showed reduced accumulation of the nuclear-encoded protein CP24 during deetiolation. Under short-day conditions reduced plastidic ATP import capacity correlates with a substantially reduced plant growth rate. This effect is absent under long-day conditions, strikingly indicating that nocturnal ATP import into chloroplasts is important. Plastidic ATP/ADP transport activity exerts significant control on lipid synthesis in developing Arabidopsis seeds. In total we made the surprising observation that plastidic ATP/ADP transport activity is not required to pass through the complete plant life cycle. However, plastidic ATP/ADP-transporter activity is required for both an undisturbed development of young tissues and a controlled cellular metabolism in mature leaves.

ATP represents the universal energy currency of all living cells. Due to both size and charge, adenylates do not cross biomembranes freely, making the involvement of highly specific transport proteins necessary. In eukaryotic cells mitochondria export ATP previously generated via oxidative phosphorylation at the matrix site in strict counter exchange to cytosolic ADP. The corresponding ADP/ATP carriers (AAC) function as dimers, comprising two identical subunits, each exhibiting six predicted transmembrane domains (Klingenberg, 1989). AAC proteins belong to the best characterized solute transporters and are the subject of numerous publications (Fiore et al., 1998).

We identified the plastidic ATP/ADP transporter as a second type of eukaryotic-adenylate carrier protein (Kampfenkel et al., 1995). Plastidic ATP/ADP transporters exist in all higher and lower plants analyzed so far (Linka et al., 2003), exhibit 11 to 12 predicted transmembrane domains (Winkler and Neuhaus, 1999), and do not show substantial structural similarities to the functional AAC homologs in mitochondria (Winkler and Neuhaus, 1999). In Arabidopsis (Arabidopsis thaliana), two plastidic ATP/ADP transporters are present, and both exhibit very similar biochemical transport properties when heterologously expressed in Escherichia coli (Möhlmann et al., 1998; Tjaden et al., 1998b).

The main function of plastidic ATP/ADP transporters is the supply of storage plastids with ATP (Schünemann et al., 1993; Kang and Rawsthorne, 1994; Neuhaus and Emes, 2000). In potato (Solanum tuberosum) tubers, the plastidic ATP/ADP transporter exerts significant control on starch accumulation (Tjaden et al., 1998a), leading to a high-flux control coefficient within this metabolic pathway (Geigenberger et al., 2001). In contrast, a recently made metabolite flux analysis on developing rapeseed (Brassica napus) embryos indicated that ATP import into corresponding plastids is not required to achieve high rates of lipid biosynthesis (Schwender et al., 2004).

All orthologs of the plastidic ATP/ADP transporter, e.g. the two isoforms from Arabidopsis, a potato ortholog, or an ortholog from the primitive red alga Galderia sulfuralia, exhibit similar transport properties in respect to substrate specificity and substrate affinity (Möhlmann et al., 1998; Tjaden et al., 1998a, 1998b; Linka et al., 2003). Therefore, the observation of similar transport properties and the contradictory information on the involvement of plastidic ATP/ADP transporters in plastidic storage product synthesis (Tjaden et al., 1998a; Schwender et al., 2004) provoke to identify reasons for the presence of plastidic nucleotide transporter (NTT) isoforms in plants and to examine the physiological impact of these ATP transporters. The analysis of the physiological implication of plastidic ATP/ADP transport activity is further encouraged since reduction of another eukaryotic ATP-transporter activity, namely the mitochondrial AAC protein, correlates with a massive, partly complete inhibition of cellular metabolism (Kolarov et al., 1990; Drgon et al., 1991) and is causative for various severe diseases (Fiore et al., 1998).

To answer the questions of expression patterns and physiological implications we started a comprehensive approach and generated in total six independent transgenic Arabidopsis plants, namely AtNTT1- and AtNTT2-promoter::GUS lines, AtNTT1::T-DNA-, AtNTT2::T-DNA-, AtNTT1-2::T-DNA-, and RNA interference (RNAi) lines. A detailed molecular analysis revealed that AtNTT2 represents a widely expressed Arabidopsis gene, whereas AtNTT1 exhibits a spatial-expression pattern. The absence of AtNTT2 or the simultaneous absence of both transporters strongly affects plant development as revealed by analysis of root formation, chloroplast maturation, and plant growth rate. Inhibition of plastidic ATP/ADP-transporter activity also exerted substantial effect on lipid content in Arabidopsis seeds. Obviously, the continuous ATP supply into developing plastids from both young seedlings and embryo tissues and into mature chloroplast at night is required for a controlled plant development.

Expression Analysis of AtNTT1 and AtNTT2

Arabidopsis possesses two isoforms of the plastidic ATP/ ADP transporter with similar biochemical transport properties. To reveal whether the presence of two independent transporter genes correlates with an organ- or development-specific expression pattern, we analyzed both the relative mRNA accumulation by northern-blot analysis and the promoter activity in transgenic plants carrying corresponding promoter--glucuronidase (GUS) fusions.

For reliable northern-blot analysis of isoform-specific mRNA accumulation it is required to use gene-specific probes. We generated probes specific for either AtNTT1- or AtNTT2 mRNA by using corresponding 3'-untranslated cDNA fragments. Although there is some minor cross hybridization, the probes used exhibited a sufficiently high specificity.

AtNTT1 mRNA accumulated strongest in root and stem tissue, and less in source leaves. In flowers and siliques the level of AtNTT1 mRNA was below or close, respectively, to the detection level. In contrast, AtNTT2 mRNA accumulated to similar amounts in roots, leaves, stem, and flower tissue. Similar to AtNTT1, the AtNTT2 mRNA was much less present in siliques.

To gain first evidence on the expression pattern of both plastidic ATP/ADP-transporter genes during early germination, we monitored the relative mRNA abundance within the first 6 d of development. Within this time span Arabidopsis develops a primary root and gains photosynthetic competence as revealed by accumulation of both chlorophyll and chlorophylla/b-binding protein (CAB) mRNA. The level of AtNTT1 mRNA within this period of development remained close to the detection level without any substantial changes. This is different from the expression of AtNTT2, as latter mRNA strongly accumulated in days 1 and 2, and declined from day 3 to a level still above the AtNTT1 mRNA.

To reveal whether plastidic ATP/ADP-transporter gene expression responds on altered sugar availabilities, we floated source leaf discs in either water (control) or 100 mM Glc or Suc.

The incubation of leaf discs for 24 h in water did not alter the levels of AtNTT1 or AtNTT2 mRNA. In contrast, the presence of Glc or Suc strongly increased the accumulation of AtNTT1 mRNA but did not influence AtNTT2 mRNA concentration.

To perform a second independent approach to study regulation of gene expression, we generated transgenic plants harboring either an AtNTT1-promoter::GUS- or AtNTT2-promoter::GUS gene, respectively. During the first 6 d of development the AtNTT1 promoter is hardly active. At day 2 AtNTT1 promoter activity is slightly detectable in the center of the primary root, and at day 6 cells comprising the vascular structures in photosynthesizing cotyledons exhibited AtNTT1 promoter activities. In contrast to this, AtNTT2 promoter activity is very high at day 1, especially in the root tip and developing cotyledons. At day 3, highest AtNTT2 promoter activity is detectable in the root hair zone and at the basis of cotyledons. At day 6, we still observed strong AtNTT2 promoter activity in the root and in rapidly developing secondary leaves, still representing strong sinks.

In source leaves, both promoters are barely active. AtNTT1 promoter-GUS activity is detectable only in the vascular bundles located at the edge of the leaf, whereas AtNTT2 promoter activity was hardly detectable in the leaf. This result does not necessary contradict the northern-blot analysis, as latter reflect the sum of AtNTT mRNA in total leaf tissue. In both flower tissue and developing siliques the AtNTT1 promoter activity is below the detection level. Petal crown leaves showed slight AtNTT2 promoter activity, which was, however similar to AtNTT1 promoter activity, nearly absent in developing siliques.

Generation of Arabidopsis Mutants Exhibiting Reduced or Abolished Plastidic ATP/ADP-Transporter Gene Expression

In the SALK library we identified a putative AtNTT1 knockout line exhibiting the T-DNA insertion in exon 1. Corresponding heterozygous plants have been selfed to obtain homozygous mutants. By use of the gene-specific primers NTT1/1 and NTT1/2, we were able to amplify a PCR product of the expected size (about 2.4 kb) on wild-type DNA but not on DNA from AtNTT1::T-DNA plants. The PCR product amplified on wild-type DNA has been sequenced to confirm the correct nucleotide sequence (data not shown). Using the primers NTT1/2 and left boarder (LB) we amplified a PCR product on DNA obtained from mutant plants but not from wild-type plants. The PCR product has been sequenced to confirm the insertion site. To check that the T-DNA insertion into the AtNTT1 gene correlates with absence of the corresponding mRNA, we performed a reverse transcription (RT)-PCR analysis. As expected, we were able to demonstrate the presence AtNTT1 mRNA in wild-type leaf tissue but not in leaves from AtNTT1::T-DNA plants. Similarly, a northern-blot analysis demonstrated the presence of AtNTT1 mRNA in wild-type leaves but not in AtNTT1::T-DNA leaves. Remarkably, the absence of AtNTT1 mRNA is not compensated by an increase of AtNTT2 mRNA.

A putative AtNTT2 knockout line was available in the GARLIC library carrying the T-DNA insertion in exon 2. Heterozygous plants have been grown and selfed to obtain a homozygous knockout line. The combination of the gene-specific primers NTT2/2 and NTT2/4 allowed amplification of a PCR product of the expected size (about 2.4 kb) on genomic wild-type DNA but not on DNA from homozygous AtNTT2::T-DNA plants.

The use of the gene-specific primer NTT2/2 and the LB primer allowed amplification of a fragment of about 1.3 kb on genomic DNA from AtNTT2:T-DNA plants but not in wild-type DNA. The PCR products have been sequenced to demonstrate that the correct DNA fragments have been amplified and to confirm the position of the T-DNA insertion (data not shown). To prove that the T-DNA insertion in the AtNTT2 gene correlates with the absence of the corresponding mRNA, we performed RT-PCR and northern-blot analysis. The gene-specific primers NTT2/2 and NTT2/4 allowed to amplify a PCR product of the expected size on cDNA prepared from wild-type leaf tissue but not on cDNA prepared from AtNTT2::T-DNA plants. Similar to this, the northern-blot analysis revealed the absence of AtNTT2 mRNA in the homozygous knockout plants but showed the presence of this mRNA in wild-type leaf tissue.

Although both genes, AtNTT1 and AtNTT2, reside on chromosome 1, we crossed homozygous AtNTT1 and AtNTT2::T-DNA lines to receive null mutants (AtNTT1-2::T-DNA), lacking both functional plastidic ATP/ADP-transporter genes. Due to the relatively wide distance of both genes on chromosome 1, it appeared likely to receive null mutants due to crossover during meiosis. We screened about 100 independent plants and identified 5 plants lacking intact genes from both transporters. That these plants represent homozygous null mutants has been demonstrated by PCR on genomic DNA. The primer combinations NTT1/1 and NTT1/2, and NTT2/2 and NTT2/4, respectively, allowed amplification of expected PCR products on genomic DNA from wild-type but not from AtNTT1-2::T-DNA plants. The use of the primer combinations NTT1/2 and LB, and NTT2/2 and LB, in contrast, allowed amplification of expected DNA fragments on DNA from homozygous AtNTT1-2::T-DNA plants but not on DNA isolated from wild-type plants.

To prove that putative effects connected with the absence of functional AtNTT1 or AtNTT2 genes, or which are present in the double-knockout line are really due to reduced levels of corresponding gene products, we created further transgenic plant lines exhibiting strongly reduced levels of both mRNA species due to an RNAi effect. For this, we cloned a 418-bp fragment from AtNTT1 (corresponding to base positions 1,006-1,424 in the AtNTT1 cDNA; Kampfenkel et al., 1995) in sense and antisense orientation into the Hannibal vector. This cDNA fragment of AtNTT1 exhibits 92% sequence identity to the corresponding cDNA domain in AtNTT2 leading to the expectation that the final RNAi construct might reduce the levels of both mRNA species simultaneously. After transformation of Arabidopsis plants, we received various independent transgenic plants with strongly reduced levels of AtNTT1 and AtNTT2 mRNA. Especially in lines 9, 10, and 14, both mRNA species were below the detection level. The absence of highly specific antisera detecting plastidic ATP/ADP transport proteins so far prevents a quantification of the final NTT protein levels in RNAi lines. However, as all RNAi lines generally exhibited physiological characteristics similar to null mutants (see below), we assume that they contained very strongly reduced transport activities, usually named knock down mutants.

Analysis of Germination and Growth Pattern of Wild-Type and Transgenic Arabidopsis Plants

After 1 d of germination, Arabidopsis shows a primary root of about 3 to 4 mm length, exhibiting a root hair zone of about 1 mm following the root tip. Mutant plants lacking the functional transporter gene AtNTT1 showed a similar size and shape of the primary root as wild-type roots. In contrast, mutant plants lacking a functional AtNTT2 gene and RNAi line 10 exhibited less developed primary roots. RNAi lines 9 and 14 showed similarly reduced roots. After 1 d of germination, AtNTT2::T-DNA plants and RNAi lines showed a substantially reduced number of rooted seedlings when compared to wild-type or AtNTT2::T-DNA plants.

To reveal whether plastidic ATP/ADP transporters are important for development of photosynthetically competent chloroplasts we analyzed chlorophyll accumulation within the first days of development. For this we germinated wild-type and mutant seeds for 5 d in a growth chamber under short-day conditions and analyzed the resulting chlorophyll content. Wild-type and AtNTT1::T-DNA seedlings showed similar levels of chlorophyll. In contrast to this, knockout plants lacking a functional AtNTT2 gene; RNAi lines 10, 9, 14; and the null mutant showed a reduced seedling size and a strongly reduced average chlorophyll content. Wild-type plants contained about 0.33 μg chlorophyll/plant, whereas chlorophyll in AtNTT2::T-DNA plants amounted to only 0.20 μg/plant. Both plants from RNAi line 10 and null mutants contained less than one-sixth of the chlorophyll present in wild-type seedlings.

To reveal whether the reduced chlorophyll level observed in some mutant lines is due to an impaired chlorophyll biosynthesis per se or might also correlate with alterations of the whole thylakoid system we examined the chloroplast ultrastructure by transmission electron microscopy. The ultrastructure of chloroplasts in 5-d-old wild-type plants exhibits a well-organized intraorganell membrane system, comprising grana and stroma thylakoids. In contrast, low-chlorophyll-containing chloroplasts from RNAi line 10 exhibited less thylakoids; especially the number of grana stacks appeared to be strongly reduced in this mutant.

To complete our picture on the effects of altered NTT gene expression on chloroplast development, we additionally analyzed the accumulation of nuclear-encoded chloroplast protein during initiation of deetiolation. For this, we germinated wild-type, RNAi, and null mutant seedlings for 6 d in the dark and illuminated etiolated seedlings for 8 or 24 h (at 100 μmol quanta m2 s-1). Subsequently, the change in the chlorophyll content was monitored, and the accumulation of chlorophyll-binding protein CP24, as indicator for altered plastidic protein import/maturation capacity, was examined by western-blot analysis.

After 6 d of dark incubation, the chlorophyll levels in wild-type and all mutant seedlings were similarly low and amounted to less than 0.01 mg/plant. After 8 and 24 h of illumination the chlorophyll level in wild-type leaves increased already to about 0.045 mg/g fresh weight (FW) and 0.160 mg/g FW, respectively. In contrast, RNAi lines 10, 9, 14, and the null mutant showed much less chlorophyll accumulation, amounting to only about 0.050 mg/g FW after 24 h of illumination. This reveals that also during the sudden induction of deetiolation, mutant plants showed a reduced capacity for chlorophyll synthesis.

In none of the Arabidopsis lines analyzed was CP24 detectable after 6 d of dark germination. However, appreciable levels of CP24 were present in wild-type seedlings after already 8 h of light induction. CP24 levels in corresponding RNAi tissue (lines 10, 9, and 14) and null mutants were significantly lower than in wild-type tissue. After 24 h of light incubation, CP24 levels were high in wild-type tissue and still significantly lower in RNAi- (lines 10, 9, and 14) and null mutants.

Besides the direct import of ATP via a plastidic ATP/ADP transporter, plastids may regenerate endogenous ATP via glycolytic enzyme activities. To raise evidence on an up-regulation of plastidic glycolytic activity during deetiolation in mutants, we quantified the level of mRNAs encoding enzymes and transporters involved. For this we germinated wild-type or RNAi plants for 6 d in the dark and illuminated subsequently etiolated seedlings for 8 h before cDNA was prepared. Gene-specific primers were chosen to amplify about 500-bp fragments coding for either plastidic phosphoglycerate kinases 1 and 2; pyruvate kinases 1, 2, and 3; or plastidic triose P/P, Glc 6-P/P, phosphoenolpyruvate/P transporters 1 and 2, and xyluose 5-P/P transporter. We observed increased mRNA levels of plastidic PGK1, PK1, and PK3 in RNAi plants compared to wild-type plants, whereas the mRNA levels of the PGK2, PK2, and all plastidic phosphate transporters have not been changed substantially in mutant tissues.

Wild-type and AtNTT1::T-DNA plants grown for 50 d under short-day conditions exhibited an average rosette size of about 12 cm. AtNTT2::T-DNA plants, RNAi, or null mutants showed, however, a strongly reduced average size of the leaf rosette, approaching only 6 and 3 cm on average. Interestingly, under long-day conditions (16 h light/d), the growth difference between wild-type plants and null mutants is nearly abolished.

The observation that RNAi and null mutants exhibited severely impaired growth tempted us to study physiological and morphological changes in these mutants in more detail. As the impaired growth is due to processes connected to a reduced plastidic ATP supply under conditions of long-night phases we first focused on changes in starch levels at the end of the day and night phase. However, we did not observe specific changes in transitory starch metabolism in AtNTT1-; AtNTT2::T-DNA; RNAi lines 10, 9, and 14; or in null mutants when compared to wild-type leaves (plants were grown under short-day conditions). All plant lines exhibited starch contents equivalent to about 30 μmol C6/mg chlorophyll at the end of the day and about 7.5 μmol C6/mg chlorophyll at the end of the night.

Seed Quality Produced by Arabidopsis Plants with Reduced Plastidic ATP/ADP-Transporter Activity

To analyze the effect of reduced plastidic ATP/ADP-transporter activity on seed quality, we grew Arabidopsis wild-type and mutant plants under long-day conditions, a light period required for induction of flowering. Fully developed seeds from wild type, AtNTT1- and AtNTT2::T-DNA-, RNAi, and AtNTT1-2::T-DNA mutants were collected from opened siliques, and the seed weight, the lipid content, and the protein levels were quantified.

Wild-type and AtNTT1::T-DNA seeds exhibited an average weight of 23 μg/seed. AtNTT2::T-DNA, RNAi, and AtNTT1-2::T-DNA seeds exhibited reduced average weights leading to 19, 20, and 18.5 μg/seed, respectively. Lipids represent the main storage product in Arabidopsis seeds. Both wild-type and AtNTT1::T-DNA seeds accumulated similar levels of storage lipids amounting to about 7.2 to 7.5 μg lipid/seed. In contrast, AtNTT2::T-DNA and RNAi seeds exhibited only 5.8 and 6.0 μg lipid/seed, respectively, and seeds from AtNTT1-2::T-DNA plants still showed only 4.5 μg lipid/seed. The protein in wild-type and AtNTT1::T-DNA seeds has been estimated to be about 4.8 μg/seed. AtNTT2::T-DNA, RNAi, and null mutants showed reduced protein levels approaching 3.3, 3.9, and 3.8 μg/seed, respectively.

Regulation of NTT Isoform Expression

ATP represents a uniquely important cellular-energy source and is required in most cell compartments to energize a wide number of anabolic and catabolic reactions. Up to now three structurally unrelated types of intracellular ATP transporters have been identified, namely the mitochondrial AAC proteins, the plastidic ATP/ADP-transporters NTT, and the peroxisomal ATP/AMP carrier found in yeast (Saccharomyces cerevisiae; Fiore et al., 1998; Winkler and Neuhaus, 1999; Palmieri et al., 2002).

The observation that the Arabidopsis genome encodes two isoforms of plastidic ATP/ADP transporters with very similar biochemical properties (Möhlmann et al., 1998) is not surprising given that even in the unicellular bakers' yeast three AAC isoforms exist (Kalorov et al., 1990). It is assumed that AAC isoforms in all eukaryotic cells occur to allow an optimal adaptation of mitochondrial ATP transport activity on changing environmental and developmental conditions (Kalorov et al.,

1990; Fiore et al., 1998). Obviously, the same holds true for the two plastidic ATP/ADP-transporter isoforms, as they also exhibit a spatially and developmentally regulated expression pattern.

AtNTT1 and AtNTT2 mRNA accumulate in photosynthesizing leaf and stem cells. We assume that the nocturnal ATP import into the chloroplasts is the main reason for expression of both plastidic ATP/ADP-transporter genes in photosynthesizing cells.

This assumption is strengthened for the following reasons: (1) Heldt (1969) demonstrated already 35 years ago that the biochemical properties of the chloroplastic ATP/ADP transporter prevents ATP export into the cytosol but allows ATP import during the night phase; and (2) the growth differences of null mutants grown under either short- or long-day conditions strikingly demonstrate that nocturnal ATP uptake into the chloroplasts is required for proper plant development. This ATP is, however, in Arabidopsis not required for degradation of transitory starch but for other, still unidentified processes.

The strong accumulation of AtNTT1 mRNA in leaf discs incubated on high Glc or Suc concentrations indicates that this gene belongs to a large group of sugar up-regulated genes (Koch, 1996) required to reprogram chloroplasts into starch-accumulating ATP-importing storage plastids. In this respect AtNTT1 is, beside the plastidic Glc6P/Pi transporter (Quick et al., 1995), a second plastid envelope transporter protein up-regulated upon sugar application.

The specific need for ATP supply into heterotrophic plastids is further indicated by the demonstration of AtNTT2 promoter activity in rapidly developing root tips and cotyledons and in petals.

This is indicative of a high ATP import into corresponding plastids, which is substantiated by the demonstration of ATP uptake and the presence of several ATP-dependent anabolic reactions in isolated chromoplasts, root plastids, and premature chloroplasts (Robinson and Wiskich, 1977; Kleinig and Liedvogel, 1980; Kleppinger-Sparace et al., 1992).

Why Does Reduced ATP Supply into Developing Plastids Impair Leaf and Root Development?

Both the northern-blot analysis and the promoter-GUS analysis indicate that especially AtNTT2 expression is high in root tips and cotyledons of developing seedlings. This observation tempted us to study the effect of altered plastidic ATP/ADP-transporter activity on both root formation and establishment of photosynthetic competence. The deletion of a functional AtNTT1 gene in Arabidopsis does not result in an impaired root formation of young seedlings, nor did it appear that chlorophyll accumulation or seedling development was affected. This observation nicely correlates with the relatively low expression of AtNTT1 in corresponding tissues. In strong contrast, the absence of a functional AtNTT2 gene or the reduction of both mRNA species (AtNTT1 and AtNTT2) in RNAi mutants led to a strongly decreased formation of primary roots in young seedlingsand a retarded chlorophyll accumulation corresponding to a reduced growth rate.

The impaired root development is most likely due to an inhibited rate of fatty-acid synthesis. In plants this process takes place exclusively in plastids, and in case of root plastids the process has been characterized to be strictly dependent upon ATP import rather than on internal ATP regeneration via glycolytic reactions (Kleppinger-Sparace et al., 1992). The localization of AtNTT2 expression in the root tip, representing the meristematic zone of cell division and elongation, is therefore fully consistent with a high demand for fatty-acid synthesis in this tissue. In addition, heterotrophic plastids are known as a cellular site for energy-consuming amino acid biosynthesis (Neuhaus and Emes, 2000). Therefore, reduced rates of amino acid synthesis in mutant lines might also contribute to an impaired root development.

In the case of developing leaf tissue several processes are negatively affected by reduced plastidic ATP/ADP-transporter activity. First, the accumulation of chlorophyll is delayed in AtNTT2::T-DNA-, AtNTT1-2::T-DNA, and RNAi seedlings. Second, the generation of functional thylakoid structures is impaired in plants with strongly reduced plastidic ATP import capabilities; and third, the accumulation of nuclear-encoded proteins in developing-mutant chloroplasts is reduced. Both chlorophyll synthesis and protein import are dependent upon the presence of ATP at the stromal site (Soll and Tien, 1998; Buchanan et al., 2000), and the inhibition of both processes in mutant plastids strikingly show that alternative routes for ATP regeneration do not compensate for insufficient import capacity. Interestingly, we did not observe an accumulation of CP24 preprotein in mutant tissues exhibiting impaired accumulation of the mature CP24 protein. This observation might indicate that a so-far unknown signaling exists between the developing chloroplast and the nucleus regulating the expression of genes encoding for chloroplastic proteins. It should be mentioned here that we were not able to compare altered NTT mRNA accumulation in transgenic mutant lines with alterations of corresponding transport protein levels. For such analysis we will attempt to generate isoform-specific antisera in the near future.

Reduced ATP Supply into Developing Seed Plastids Limits Lipid Accumulation

We showed in the past that starch accumulation in potato tubers is strongly affected by altering the plastidic ATP/ADP-transporter activity (Tjaden et al., 1998a) leading to a high metabolic-flux control coefficient (Geigenberger et al., 2001). Therefore, we analyzed whether reduced plastidic ATP import capacity governs the end-product accumulation in Arabidopsis embryos to a similar extent as observed in potato. This analysis was further encouraged, since experiments on isolated rapeseed seed-embryo plastids showed that the highest rates of fatty-acid synthesis depend upon the supply with exogenous ATP (Eastmond and Rawsthorne, 1998; Rawsthorne, 2002), whereas a recently developed mathematical carbon-flux model indicated that net ATP import is not required for maximal fatty-acid synthesis in rapeseed embryos (Schwender et al., 2004).

As given, AtNTT1::T-DNA did not show altered seed weight, lipid, and protein content when compared to wild-type seeds, whereas AtNTT2::T-DNA seeds showed reduced weight, which correlates with reduced levels of lipids and storage protein. Strongest reduction of the lipid content showed seeds generated from double-knockout mutants as these seeds contained only about 50% of the lipid content present in wild-type seeds. This result is surprising, because the expression level of NTT1 and NTT2 mRNA in developing siliques and seeds is remarkable low. Obviously, even low mRNA levels allow the maintenance of sufficient plastidic ATP import capacity.

It is important to note that the reduced seed oil phenotype is evident under long-day conditions, where the effects of gene knockout on whole-plant physiology, and hence maternal carbon supply to the embryo, were absent. These effects on storage product content are therefore likely to be specific to alterations to NNT gene expression in the seed. From this result we conclude that, similar to rapeseed and cauliflower (Brassica oleracea) inflorescence plastids (Möhlmann et al., 1994; Eastmond and Rawsthorne, 1998), Arabidopsis embryo plastids need to import cytosolic ATP to achieve the highest rates of lipid synthesis. This conclusion is fully consistent with recent observations that the overall energy status of developing rapeseed seeds correlate with lipid synthesis (Vigeolas et al., 2003).

However, Arabidopsis embryo plastids obviously possess, in addition to ATP/ADP-transporter proteins, endogenous sources for ATP regeneration because the absence of both transporter activities does not correlate with a total loss of storage product. In general two other metabolic pathways might allow stromal regeneration of ATP: First, chlorophyll-containing embryo plastids might regenerate ATP by photophosphorylation. Secondly, stromal-located glycolytic sequences might regenerate ATP at the enzymic steps catalyzed by phosphoglycerate kinase (PGK), or pyruvate kinase (PK). We would like to exclude the first possibility, since in case of rapeseed the light transmission into the developing seed tissue is supposed to be too low (Eastmond et al., 1996). Moreover, rapeseed mutants showing strongly reduced chlorophyll levels in developing embryos did not contain less lipids than wild-type plants (Tsang et al., 2003).

These independent observations point to stromal glycolysis as the alternative source for endogenous ATP resynthesis. This assumption is substantiated by the demonstration that Glc 6-phosphate is a very suitable carbon precursor for fatty-acid synthesis in rapeseed plastids (Kang and Rawsthorne, 1996). In addition, the observation that developing leaf plastids from RNAi plants exhibited increased accumulation of plastidial PGK- and PK mRNAs during deetiolation might indicate that heterotrophic Arabidopsis plastids use endogenous glycolysis for ATP resynthesis.

The latter observation is moreover in full agreement with the demonstration of a DHAP-driven ATP production in etioplasts from dark-grown barley leaves (Batz et al., 1992). Obviously, some types of heterotrophic plastids use endogenous glycolysis for stromal ATP production, which contribute to energize anabolic reactions and compensate partly the lack of plastidic ATP import capacity in mutant plants. Moreover, the observation that mRNA coding for plastidic glycolytic enzymes specifically accumulated in deetiolating plastids from RNAi plants might indicate that the stromal ATP (energy) status is sensed and governs expression of genes allowing regeneration of ATP by alternative sources.

CONCLUSION

Arabidopsis contains two isoforms of plastidic ATP/ADP transporter to allow an optimal spatially and developmentally regulated adaptation of gene expression. Surprisingly, Arabidopsis does not need plastidic ATP/ADP-transporter activity to pass through the complete developmental cycle. However, plastidic ATP/ADP-transporter activity is required for a controlled development of young tissues, especially shown for roots and cotyledons, and is required in mature chloroplasts at night.

The absence of plastidic ATP import in developing embryo tissue correlates with a reduction of lipid accumulation, which however still occurs at appreciable levels. This observation points to an ATP regeneration by stromal-located glycolytic enzymes, which seems to participate on ATP provision.

MATERIALS AND METHODS

AtNTT1 (AtNTT1::T-DNA) and AtNTT2 (AtNTT2::T-DNA) Knockout Mutant Plants

The heterozygous AtNTT1::T-DNA mutant plant (Salk_013530) was provided by the SALK library. In that mutant the T-DNA is located in the first exon of AtNTT1 (locus At1g80300) on bp position 777. The heterozygous AtNTT2::T-DNA mutant (GARLIC_ 288_E08.b.1a.Lb3Fa) was provided by the Torrey Mesa Research Institute (San Diego). In that mutant the T-DNA is located in the second exon of AtNTT2 (locus At1g15500) on bp position 1,015.

To confirm that we generated homozygous mutants after backcrossing, we used gene- and T-DNA-specific primers. For PCR on genomic DNA the following primers were used: NTT1/1(5'-TTTCTTCTGTGTATCTGCGGGAGAGAGTG-3'); NTT1/2 (5'-CTTTCTTTCCCCCCCAACAAAACCAAATA-3'); SALK_LB (5'-ACTCAACCCTATCTCGGGCTATTC-3'); NTT2/4 (5'-TCTCTTCTCCTCTCTACCCAGAGC-3'); NTT2/2 (5'-CCAAATCCCAAAACCCTTTTATTCATC-3'); and GARLIC_LB (5'-TAGCATCTGAATTTCATAACCAATCCGATACAC-3').

Generation of Double-Knockout Plants (AtNTT1-2::T-DNA)

To generate double-knockout mutants (designated AtNTT1-2::T-DNA) lacking both functional plastidic ATP/ADP-transporter genes, homozygous AtNTT1::T-DNA and homozygous AtNTT2::T-DNA mutant plants were crossed. Although both genes reside on chromosome 1, we were able to identify double-knockout mutants by use of the primers given above.

Generation of RNAi Mutants

Transgenic RNAi plants were generated to achieve strongly reduced mRNA-levels of both AtNTT1 and AtNTT2. For Arabidopsis (Arabidopsis thaliana) transformation the pART27 vector (Gleave, 1992) was used. For this we cloned a 418-bp fragment from AtNTT1 (corresponding to bp positions 1,006-1,424) in sense and antisense orientation into the pHANNIBALL vector (Wesley et al., 2001). This cDNA sequence of AtNTT1 exhibits 92% identity to the corresponding cDNA sequence of AtNTT2. For the sense construct, the following primers were used: RNAi-sense-XhoI-fp, 5'-GATATGTTCCTCGA-GCAACCCGTA-3'; and RNAi-sense-EcoRI-rp, 5'-GGAATTCCAAGTAGCGGTGTCATACC-3'. For the antisense construct, the following primers were used: RNAi-antisense-XbaI-fp, 5'-GATATGTTCCTCTAGAAACCCGTA-3'; and RNAi-antisense-BamHI-rp, 5'-CGGGATCCCGAAGTAGCGGT-GTCATACC-3'. The sense construct was restricted with XhoI and EcoRI and the antisense construct with XbaI and BamHI. Restriction sites added by the primers ensured the correct orientation of the resulting sense and antisense constructs. The resulting pHANNIBALL constructs were subcloned as NotI fragments into pART27, and the final plasmid was subsequently transformed into Agrobacterium. Transformation of Arabidopsis was conducted according to the floral-dip method.

Generation of AtNTT1::Promoter-GUS and AtNTT2::Promoter-GUS Plants

For the generation of promoter-GUS constructs the binary vector pGPTV (Becker et al., 1992) containing the - glucuronidase-(uidA-) gene from Escherichia coli was used. For the generation of promoter-GUS fusion constructs a promoter region of about 1.4 kb of either the AtNTT1 or AtNTT2 gene was cloned upstream of the GUS gene. The promoter region of the AtNTT1 or AtNTT2 gene (including 21 bp of the coding region) was amplified by PCR from genomic DNA. Both promoters were sequenced to check that the correct products were amplified.

For amplification of the AtNTT1 promoter the following primers were used: JR1-sense, 5'-TGGACCTACATATGGG-TTCGATTCGACTCC-3'; and JR2ant, 5'-AAGAGAGAAG-CCCCCGGGTTTGAATCACAGC-3'. For amplification of the AtNTT2 promoter the following primers were used: JR3-sense, 5'-GGAAGAATCTGAAGTTTTGGAACCC-3'; and JR4-anti, 5'-GAGAGAATTCCCCGGGTTTGAATCAG-3'. After blunt-end ligation of the PCR products in T7 orientation into the SmaI-restricted pBSK vector, both promoters were restricted with HindIII and SmaI and subsequently inserted in frame with the GUS gene. The resulting constructs were used for Agrobacterium transformation. Transformation of Arabidopsis was as given above.

Generation of AtNTT1- and AtNTT2-Specific Probes for Northern-Blot Analysis

Gene-specific probes, each corresponding to the respective 3'untranslated region, were generated by PCR on cDNA. AtNTT1-specific probes were amplified with the following primers: At1oligo3sense, 5'-GGAGAAATCTGCTCC-3'; and At1oligo1anti, 5'-ACTTCAACGATACACACAAAGG-3'. AtNTT2-specific probes were amplified with the following primers: At2oligo3sense, 5'-ACTGGCATTTAGACG-3'; and At2oligo1anti, 5'-CTAGTTTGGTATTGG-3'. The PCR products were subsequently cloned into the pGEMTeasy vector. For northern-blot analysis the cloned fragments were excised, separated by gel electophoresis, gel purified, and radioactively labeled with [32P]-dCTP by random priming, using the Ready To Go kit (Amersham-Pharmacia, Freiburg, Germany).

Gene Expression Studies

Poly(A+) mRNA was isolated from rosette leaves or whole seedlings (0.1 g each) by use of Dynabeads (Dynal AS, Oslo) and converted to cDNA by reverse transcription (SuperscriptII, Invitrogen, Carlsbad, CA). For semiquantitative RT-PCR reactions were carried out with 1 μL template and 1 unit Taq-Polymerase in a total volume of 50 μL. PCR conditions were 3 min at 95°C, followed by 25 cycles of 30 s at 96°C, 30 s at 50°C, and 60 s at 72°C. Primers used for amplification are listed. Gene-specific primers were used to amplify PCR products of 450 to 550 bp.

The following primers were used: pyruvate kinase I-fp, (At1g32440) 5'-GTGCGACTCTTCCATCCATT-3'; pyruvate kinase I-rp, (At1g32440) 5'-GGTTCTCAACGCCACAGTAT-3'; pyruvate kinase II-fp, (At3g22960) 5'-CAAGGCGCTCACGG-TTCTAA-3'; pyruvate kinase-II-rp, (At3g22960) 5'-

CATGTCCGAGACTGCGATCA-3'; pyruvate kinase III-fp, (At3g52920) 5'-CTCCTGAAGATGTGCCTAAC-3'; pyruvate kinase III-rp, (At3g52920) 5'-CCAGTCCTTCTCAGTGATTG-3'; phosphoglycerate kinase I-fp, (At1g56190) 5'-ACGATGGCGAAGAAGAGTGT-3'; phosphoglycerate kinase I-rp, (At1g56190) 5'-ACCACCAAGCAGAAGGATGT-3'; phosphoglycerate kinase II-fp, (At3g12780) 5'-CTACCGAAGGAGTCACTAAG-3'; phosphoglycerate kinase II-rp, (At3g12780) 5'-CTGAGTCTCCTCCTCCTATT-3'; phosphoenolpyruvate translocator I-fp, (At5g33320) 5'-CATCTTGCCGCTTGCTGTTGT-3'; phosphoenolpyruvate translocator I-rp, (At5g33320) 5'-TGGAAGCAGAG-TGCAGCGATAA-3'; phosphoenolpyruvate translocator II-fp, (At3g01550) 5'-TCAGCGTTGAGCAGAAGAAG-3'; phosphoenolpyruvate translocator II-r, (At3g-01550) 5'-TCACCGAGCAAGAGAACAGA-3'; triose-phosphate translocator-fp, (At5g46110) 5'-CTGAAGGTGGAGA-TACCGCTG-3'; triose-phosphate translocator-rp, (At5g46110) 5'-GAGTGCGATGATGGAGATGTA-3'; Glc 6-phosphate translocator-fp, (At5g54800) 5'-TTCCATCGACGGAGCTTCCA-3'; Glc-6-phosphate translocator-rp, (At5g54800) 5'-ACGCAGGTTCACCACTCTTG-3'; xylulose-5-phosphate translocator-fp, (At5g17630) 5'-CCGTTGGCTCA-TCGGATTCAA-3'; xylulose-5-phosphate translocator-rp, (At5g17630) 5'-GCTCTGTAAGCTACGTTTAGA-3'; Actin-fp, 5'-TGTACGCCAGTGGTCGTACAACC-3'; and Actin-rp, 5'-GGAGCAAGAATGGAACCACCG-3'.

Cultivation of Plant Material

Wild-type and transgenic Arabidopsis plants (ecotype Columbia) were grown in a climate-controlled chamber on soil at 22°C and 100 μmol quanta m2 s-1. Prior to germination, seeds were incubated for 2 d in the dark at 4°C for imbibition (Weigel and Glazebrook, 2002). For short-day growing conditions, the light was given for 10 h/d; for long-day conditions light was present for 16 h/d. For root growth and seedling analysis, surface-sterilized seeds were sown on half-concentrated Murashige and Skoog (1962) plates, containing 0.8% Agar, 0.05% MES (adjusted to pH 5.7 with KOH), and 1% Suc. Prior to germination, plates were incubated at 4°C for 2 d in the dark and subsequently transferred to the growth chamber, and growth was continued for 24 h under long-day conditions.

Extraction of Total RNA and RNA Gel-Blot Hybridization

Total RNA was isolated from frozen tissue samples (liquid nitrogen) by using the Purescript extraction kit (Gentra Systems, North Minneapolis, MN), according to the manufacturer's instructions. For RNA gel-blot hybridization analysis, standard methods (Sambrook et al., 1989) were used. Blots were visualized by a Phospho-Imager (Packard, Frankfurt).

Histochemical Localization of GUS

Whole seedlings or tissue from transgenic plants were collected in glass scintillation vials, filled with ice-cold 90% acetone, and incubated for 20 min at room temperature. Subsequently, the samples were stained according to standard protocols (Weigel and Glazebrook, 2002).

Immunological Analysis

Accumulation of chlorophyll-binding protein CP24 during illumination was examined by western-blot analysis. Antibodies were kindly provided by Prof. R.B. Klösgen (Pflanzenphysiologie, Martin-Luther Universität Halle, Germany). Plant tissue (0.5 g) frozen in liquid nitrogen was homogenized in 250 µL buffer A (50 mM HEPES, 5 mM MgCl2, pH 7.5, 2% SDS, 1% Triton X-100, 15% glycerol, 1 mM EDTA, phenylmethylsulfonyl fluoride [PMSF] 1/100 [v/v]) at room temperature. SDS-PAGE, northern transfer, and immunodetection were conducted according to standard protocols.

Transmission Electron Microscopy

For chloroplast ultrastructure analysis from wild-type and RNAi mutants cotyledons from 5-d-old seedlings, grown under short-day conditions, were used. The seedlings were fixed with solution 1 (3% [v/v] glutaraldehyd, 30 mM PIPES, pH 7.0) for 1 h and subsequently washed two times for 10 min in cacodylat buffer, pH 7.0 (50 mM sodium cacodylat, 6.4 mM HCl). The samples were post fixed in solution 2 (1% [w/v] osmium tetroxide, 50 mM sodium cacodylat, 6.4 mM HCl, pH 7.0) for 1 h and washed as described above. Subsequently, samples were incubated for 1 h in 0.5% uranylacetat, followed by a serial dehydration with 30%, 50%, 70%, 90%, and 100% (v/v) of acetone in water. The specimens were infiltrated with a series of 25%, 50%, 75%, and 100% Spurr (Ted Pella, Redding, CA) in acetone. After embedding in Spurr the blocks were sectioned and stained with 2% uranyl acetate and lead citrate before viewing in a transmission electron microscope (Zeiss, Oberkochen, Germany).

Seed Analysis

For lipid quantification, 0.1 g completely mature and air-dried seeds were homogenized in a mortar in liquid nitrogen. Subsequently, 1.5 mL isopropanol was added and further homogenized. The suspension was transferred into a 1.5-mL reaction tube and incubated for 12 h at 4°C on a laboratory shaker at 100 rpm. Subsequently, samples were centrifuged at 12,000g for 10 min and the supernatant was transferred into preweighted 1.5-mL reaction tube. Tubes were incubated at 60°C for 8 h to evaporate the isopropanol. Subsequently, total lipid was quantified gravitometrically.

For seed protein quantification 0.1-g seeds were homogenized in a mortar at room temperature. Subsequently, 1,000 μL buffer medium 1 (50 mM HEPES, 5 mM MgCl2, pH 7.5, 1% Triton X-100, 15% glycerol, 2% SDS, 1 mM EDTA, PMSF, 1/100 [v/v]) was added and further homogenized. The suspension was transferred into 1.5-mL reaction tubes, and samples were centrifuged at 12,000g at room temperature for 10 min. The supernatant was transferred into new 1.5-mL reaction tubes, and proteins were quantified with bicinchoninic acid reagent (Pierce Chemical, Rockford, IL) according to manufacturer's instructions.

Bibliography

Chambers, K. L.: *Biochemical Coevolution: Eugene*, Oregon State University Press, Biology Colloquium, 1970.

Clark K. Colton: *Advances in Chemical Engineering: Research and Education*, New York, Academic Press, 1991.

David Knight: *Ideas in Chemistry: A History of the Science*, New Brunswick, N. J., Rutgers University Press, 1992.

G. M. Caroe: *William Henry Bragg, 1862–1942: Man and Scientist*, New York: Cambridge University Press, 1978.

John Hudson: *The History of Chemistry*, New York, Chapman & Hall, 1992.

Lehninger, A. L.: *Biochemistry*, New York, Worth and Co., 1970.

M. Weatherall: *In Search of a Cure: A History of Pharmaceutical Discovery*, New York, Oxford University Press, 1990.

Peter H. Spitz: *Petrochemicals: The Rise of an Industry*, New York, John Wiley & Sons, 1988.

Robert Clarke: *Ellen Swallow: The Woman Who Founded Ecology*, Chicago: Follett, 1973.

Roy Porter: *The Biographical Dictionary of Scientists*, New York, Oxford University Press, 1994.

S. T. Mossman; Peter J. T. Morris: *The Development of Plastics*, Cambridge, Royal Society of Chemistry, 1994.

Vivian Ovelton Sammons: *Blacks in Science and Medicine*, New York/London, Hemisphere Publishing, 1990.

William F. Furter: *A Century of Chemical Engineering*, New York, Plenum Press, 1982.

William H. Brock: *The Norton History of Chemistry*, New York, W. W. Norton, 1993.

Index

□□□